Natural Disaster Research, Prediction and Mitigation

Wildfires

Air Quality Impacts and Smoke Exposure

NATURAL DISASTER RESEARCH, PREDICTION AND MITIGATION

Additional books and e-books in this series can be found on Nova's website under the Series tab.

NATURAL DISASTER RESEARCH, PREDICTION AND MITIGATION

WILDFIRES

AIR QUALITY IMPACTS AND SMOKE EXPOSURE

DAVE HAWKINS
EDITOR

Copyright © 2020 by Nova Science Publishers, Inc.

All rights reserved. No part of this book may be reproduced, stored in a retrieval system or transmitted in any form or by any means: electronic, electrostatic, magnetic, tape, mechanical photocopying, recording or otherwise without the written permission of the Publisher.

We have partnered with Copyright Clearance Center to make it easy for you to obtain permissions to reuse content from this publication. Simply navigate to this publication's page on Nova's website and locate the "Get Permission" button below the title description. This button is linked directly to the title's permission page on copyright.com. Alternatively, you can visit copyright.com and search by title, ISBN, or ISSN.

For further questions about using the service on copyright.com, please contact:
Copyright Clearance Center
Phone: +1-(978) 750-8400   Fax: +1-(978) 750-4470   E-mail: info@copyright.com.

NOTICE TO THE READER

The Publisher has taken reasonable care in the preparation of this book, but makes no expressed or implied warranty of any kind and assumes no responsibility for any errors or omissions. No liability is assumed for incidental or consequential damages in connection with or arising out of information contained in this book. The Publisher shall not be liable for any special, consequential, or exemplary damages resulting, in whole or in part, from the readers' use of, or reliance upon, this material. Any parts of this book based on government reports are so indicated and copyright is claimed for those parts to the extent applicable to compilations of such works.

Independent verification should be sought for any data, advice or recommendations contained in this book. In addition, no responsibility is assumed by the Publisher for any injury and/or damage to persons or property arising from any methods, products, instructions, ideas or otherwise contained in this publication.

This publication is designed to provide accurate and authoritative information with regard to the subject matter covered herein. It is sold with the clear understanding that the Publisher is not engaged in rendering legal or any other professional services. If legal or any other expert assistance is required, the services of a competent person should be sought. FROM A DECLARATION OF PARTICIPANTS JOINTLY ADOPTED BY A COMMITTEE OF THE AMERICAN BAR ASSOCIATION AND A COMMITTEE OF PUBLISHERS.

Additional color graphics may be available in the e-book version of this book.

Library of Congress Cataloging-in-Publication Data

ISBN: 978-1-53617-182-2

Published by Nova Science Publishers, Inc. † New York

CONTENTS

Preface		**vii**
Chapter 1	Air Quality Impacts of Wildfires: Mitigation and Management Strategies *Committee on Energy and Commerce*	**1**
Chapter 2	Prepare for Fire Season *United States Environmental Protection Agency*	**335**
Chapter 3	Reduce Your Smoke Exposure *United States Environmental Protection Agency*	**339**
Chapter 4	Protect Yourself from Ash *United States Environmental Protection Agency*	**345**
Chapter 5	Protecting Children from Wildfire Smoke and Ash *United States Environmental Protection Agency*	**349**
Chapter 6	Protect Your Pets from Wildfire Smoke *United States Environmental Protection Agency*	**355**
Chapter 7	Protect Your Large Animals and Livestock from Wildfire Smoke *United States Environmental Protection Agency*	**361**
Index		**367**
Related Nova Publications		**381**

PREFACE

Chapter 1 develops a better understanding of the health impacts of wildfires and what should be done to minimize those impacts. It looks closely at the mitigation and management strategies for reducing air quality risks from wildfire smoke. In large part, these strategies involve efforts to reduce the intensity and frequency of wildfires that threaten communities. Chapters 2-7 contain U.S. Environmental Protection Agency fact sheets on how to reduce your risks to wildfire smoke and ash.

Chapter 1 - This is an edited, reformatted and augmented version of Hearing before the Subcommittee on Environment, of the Committee on Energy and Commerce, House of Representatives, One Hundred Fifteenth Congress, Second Session, Serial No. 115–165, dated September 13, 2018.

Chapter 2 - If you live in an area where the wildfire risk is high, take steps now to prepare for fire season. Being prepared for fire season is especially important for the health of children, older adults, and people with heart or lung disease.

Chapter 3 - When wildfires create smoky conditions, there are things you can do, indoors and out, to reduce your exposure to smoke. Reducing exposure is important for everyone's health — especially children, older adults, and people with heart or lung disease.

Chapter 4 - Protect yourself from harmful ash when you clean up after a wildfire. Cleanup work can expose you to ash and other products of the fire that may irritate your eyes, nose, or skin and cause coughing and other health effects. Ash inhaled deeply into lungs may cause asthma attacks and make it difficult to breathe.

Ash is made up of larger and tiny particles (dust, dirt, and soot). Ash deposited on surfaces both indoors and outdoors can be inhaled if it becomes airborne when you clean up. Ash from burned structures is generally more hazardous than forest ash.

Chapter 5 - Children are especially at risk for health effects from exposure to wildfire smoke and ash, mostly because their lungs are still growing.

Wildfire concerns include the fire itself, the smoke and ash, and the chemicals from materials that have burned, such as furniture.

Smoke can travel hundreds of miles from the source of a fire. Pay attention to local air quality reports during fire season, even if no fire is nearby.

Chapter 6 - Your pets can be affected by wildfire smoke. If you feel the effects of smoke, they probably do, too!

Smoke can irritate your pet's eyes and respiratory tract. Animals with heart or lung disease and older pets are especially at risk from smoke and should be closely watched during all periods of poor air quality.

Chapter 7 - Your animals can be affected by wildfire smoke. If you feel the effects of smoke, they probably do too! High levels of smoke are harmful. Long exposure to lower levels of smoke can also irritate animals' eyes and respiratory tract and make it hard for them to breathe. Reduce your animals' exposure to smoke the same way you reduce your own: spend less time in smoky areas and limit physical activity. Animals with heart or lung disease and older animals are especially at risk from smoke and should be closely watched during all periods of poor air quality. Take the following actions to protect your large animals and livestock against wildfire smoke.

In: Wildfires
Editor: Dave Hawkins

ISBN: 978-1-53617-182-2
© 2020 Nova Science Publishers, Inc.

Chapter 1

AIR QUALITY IMPACTS OF WILDFIRES: MITIGATION AND MANAGEMENT STRATEGIES[*]

Committee on Energy and Commerce

Thursday, September 13, 2018
House of Representatives,
Subcommittee on Environment,
Committee on Energy and Commerce,
Washington, D.C.

The subcommittee met, pursuant to call, at 1:15 p.m., in Room 2123, Rayburn House Office Building, Hon. John Shimkus, [chairman of the subcommittee] presiding.

---

[*] This is an edited, reformatted and augmented version of Hearing before the Subcommittee on Environment, of the Committee on Energy and Commerce, House of Representatives, One Hundred Fifteenth Congress, Second Session, Serial No. 115–165, dated September 13, 2018.

Present: Representatives Shimkus, McKinley, Harper, Johnson, Flores, Hudson, Walberg, Carter, Duncan, Walden (ex officio), Tonko, Ruiz, Peters, DeGette, McNerney, Cardenas, Matsui, and Pallone (ex officio).

Staff Present: Samantha Bopp, Staff Assistant; Karen Christian, General Counsel; Kelly Collins, Legislative Clerk, Energy and Environment; Wyatt Ellertson, Professional Staff, Energy and Environment; Margaret Tucker Fogarty, Staff Assistant; Theresa Gambo, Human Resources/Office Administrator; Jordan Haverly, Policy Coordinator, Environment; Mary Martin, Chief Counsel, Energy and Environment; Sarah Matthews, Press Secretary, Energy and Environment; Drew McDowell, Executive Assistant; Brannon Rains, Staff Assistant; Peter Spencer, Senior Professional Staff Member, Energy; Austin Stonebraker, Press Assistant; Hamlin Wade, Special Advisor, External Affairs; Everett Winnick, Director of Information Technology; Jean Fruci, Minority Energy and Environment Policy Advisor; Caitlin Haberman, Minority Professional Staff Member; Rick Kessler, Minority Senior Advisor and Staff Director, Energy and Environment; Alexander Ratner, Minority Policy Analyst; and Catherine Zander, Minority Environment Fellow.

OPENING STATEMENT OF HON. JOHN SHIMKUS, A REPRESENTATIVE IN CONGRESS FROM THE STATE OF ILLINOIS

Mr. Shimkus. I am going to call the committee to order and make a brief statement before I give my opening statement, is that we will have the chairman and the ranking member both come in their due time, and then we will break and allow them to do their opening statements. At least we can get started on time, if that is agreeable with everybody, which it seems like it is.

I now recognize myself 5 minutes for an opening statement.

A year ago, we took testimony to examine the air quality impacts of wildfires with the focus on stakeholder perspectives. Given the community's jurisdiction over air quality policies and public health, the goal then, as it is today, was to develop a better understanding of the health impacts of wildfires and what should be done to minimize those impacts.

We return to the topic this afternoon to look closely at the mitigation and management strategies for reducing air quality risks from wildfire smoke. In large part, these strategies involve efforts to reduce the intensity and frequency of wildfires that threaten communities.

The strategies also involve managing the inevitable smoke impacts, whether from wildfires or from what is known as prescribed burning. And they involve ensuring that effective actions are credited appropriately in air quality planning, air quality monitoring, and compliance activities, so States and localities are not punished for taking action that will improve public health.

Last year, some 10 million acres were burned in the United States by wildfires, the second worst fire season since 1960. As of last week, this fire season has resulted in more than 7 million acres burned, with acute impacts of smoke lingering for extended periods of time throughout California and the Pacific Northwest.

The urgency for reducing the severity of these fires is underscored by news reports and reports from this committee's own members, including Chairman Walden, of the impacts of wildfire smoke. This smoke can smother communities with high levels of particulate matter and other respiratory irritants.

These levels, which are manyfold over normal air quality, intensify asthma and chronic pulmonary diseases, and impact the lives of millions of people.

Against this backdrop are a panel of witnesses who can speak to the complex set of strategies that are needed to more effectively address wildfires and smoke risk.

We will hear today from two State foresters who oversee and implement fire management strategies in their States: Sonya Germann from the Montana Department of Natural Resources and Conservation, Tom – I hope this – Boggus.

Mr. Boggus. Boggus.

Mr. Shimkus. Boggus. Thank you. Bogus was a word we used at West Point. Boggus is better, so – the Texas State forester and director of Texas A&M Forest Service.

While the general approaches among State forestry officials to mitigating risks are consistent, there are regional differences that affect what is put into practice and can inform future policymakers.

We will hear a State air quality perspective. Mary Anderson, who is with the Idaho Department of Environmental Quality, can help us understand the practical challenges of managing wildfire smoke and how her agency works to address air quality risks.

Collin O'Mara, president of the National Wildfire Federation, has been before the committee before, brings an environmental perspective, but is also experienced as a former head of the State of Delaware's Department of Natural Resources and Environment Control.

And finally, we will hear from Oregon State Senator Herman Baertschiger from southern Oregon, who has extensive experience in forestry and wildland firefighting. I am looking forward to his perspective on what to do and his perspective on the impacts of wildfires on his constituents.

Let me welcome the panelists. I look forward to understanding the challenges and the opportunities you face and what you can do to ensure our Federal air regulations accommodate these strategies.

And with my remainder of time, I would like to yield to the gentleman of Texas, Mr. Flores.

[The prepared statement of Mr. Shimkus follows:]

OPENING STATEMENT OF THE HONORABLE JOHN SHIMKUS CHAIRMAN, SUBCOMMITTEE ON ENVIRONMENT, AIR QUALITY IMPACTS OF WILDFIRES: MITIGATION AND MANAGEMENT STRATEGIES SEPTEMBER 13, 2018

A year ago, we took testimony to examine the air quality impacts of wildfires, with a focus on stake holder perspectives. Given the Committee's jurisdiction over air quality policies and public health, the goal then, as it is today, was to develop a better understanding of the health impacts of wildfires and what should be done to minimize those impacts.

We return to the topic this afternoon, to look closer at the mitigation and management strategies for reducing the air quality risks from wildfire smoke. In large part, these strategies involve efforts to reduce the intensity and frequency of wildfires that threaten communities.

The strategies also involve managing the inevitable smoke impacts, whether from wildfires or from what is known as prescribed burning. And they involve ensuring that effective actions are credited appropriately in air quality planning, air quality monitoring, and compliance activities, so states and localities are not punished for taking action that will improve public health.

Last year, some 10 million acres were burned in the United States by wildfires, the second worst fire season since 1960. As of last week, this fire season has resulted in more than 7 million acres burned, with acute impacts of smoke lingering for extended periods of time, throughout California and the Pacific Northwest.

The urgency for reducing the severity of these fires is underscored by news reports – and reports from this Committee's own members, including Chairman Walden — of the impacts of wildfire smoke. This smoke can smother communities with high levels of particulate matter and other respiratory irritants. These levels, which are many-fold over normal air quality, intensify asthma and chronic pulmonary diseases, and impact the daily lives of millions of people.

Against this backdrop, our panel of witnesses can speak to the complex set of strategies that are needed to more effectively address wildfires and smoke risks.

We will hear today from two state foresters, who oversee and implement fire management strategies in their states. Sonya Germann, from the Montana Department of Natural Resources and Conservation and Tom Boggus, the Texas State Forester and Director of the Texas A&M Forest Service. While the general approaches among state forestry officials to mitigating risks are consistent, there are regional differences that affect what is put into practice and can inform future policymaking.

We will hear a state air quality perspective. Mary Anderson, who is with the Idaho Department of Environmental Quality, can help us understand the practical challenges of managing wildfire smoke, and how her agency works to address air quality risks.

Collin O'Mara, President of the National Wildlife Federation, brings an environmental perspective but also experience as the former head of the State of Delaware's Department of Natural Resources and Environmental Control.

And finally, we will hear from Oregon State Senator Herman Baertshiger (BEAR CHIGGER), from southern Oregon, who has extensive experience in forestry and wildland firefighting. I'm looking forward to his perspective on what to do, and his perspective on the impacts of wildfires on his constituents.

Let me welcome the panelists. I look forward to understanding the challenges and opportunities you face, and what we can do to ensure our federal air regulations accommodate these strategies.

Mr. Flores. So thank you, Mr. Chairman, for yielding me a part of your time, and thank you for holding today's important hearing.

I am pleased to welcome my constituent, Mr. Tom Boggus, to today's hearing. He is testifying on behalf of the National Association of State Foresters. Mr. Boggus is a native of Fort Stockton, Texas, and he joined the Texas A&M Forest Service in 1980. He was appointed as the director and State forester of the Texas A&M Forest Service in February of 2010, and he has extensive familiarity with the issue we are going to be discussing today.

I look forward to hearing from him, along with the rest of our expert witnesses, on how we can appropriately manage our forests to minimize wildfire impacts.

Thank you, and I yield back.

Welcome, Mr. Boggus.

Mr. Shimkus. The gentleman yields back his time.

The chair now recognizes the ranking member of the subcommittee, Mr. Tonko, for 5 minutes.

OPENING STATEMENT OF HON. PAUL TONKO, A REPRESENTATIVE IN CONGRESS FROM THE STATE OF NEW YORK

Mr. Tonko. Thank you, Mr. Chair. And thank you to our witnesses for being here this afternoon.

As some of you may remember, this subcommittee held a similar hearing last year on wildfires and air quality issues. Since that time, we confirmed that, in 2017, more than 66,000 wildfires burned approximately 10 million acres. 2018 is proving to be another difficult year. Right now, there are over 80 active fires covering over a million acres and threatening people's health and safety and property, including the Mendocino Complex fire, the largest recorded fire in California's history.

Undeniably, these fires have become increasingly worse in recent years. Today, we will hear about the consequences of wildfires to both human health as well as forest health. Smoke, which includes particulate matter, is harming people, and the growing number and size of these fires are erasing the gains that have been made under the Clean Air Act in reducing fine particulate matter pollution.

We will also hear about the best practices in forest management, including prescribed burns and other tools that can mitigate some of the worst impacts of these fires and reduce the harm of smoke. While I do not follow this issue as closely as many of our western colleagues, my understanding is that historically the method for funding the United States Forest Service emergency fire response has been a major factor in limiting funding for more proactive forest management activities.

In March, Congress passed the fiscal year 2018 omnibus appropriations bill, which included a fire funding fix that will take effect in fiscal year 2020. I acknowledge that more may need to be done to promote better forest management techniques, but we must see how this fix plays out before adopting new major provisions that undermine environmental laws in our national forests.

As we discuss the devastation that can be caused by Mother Nature, we must also acknowledge our fellow Americans that are facing down Hurricane Florence. Whether it is hurricanes on the East Coast or fires out west, we are experiencing more frequent and costly natural disasters across our country. As with hurricanes, climate change creates conditions that make wildfires worse. Droughts, dryer soils, and higher temperatures, all associated with climate change, are resulting in a longer fire season and causing an increase in the severity and frequency of wildfires.

A 2016 study published in the Proceedings of the National Academy of Sciences concluded that human-caused climate change is responsible for the doubling of the area burned by wildfires since 1984. In 2017, the National Wildfire Federation, which is represented here today by NWF President Collin O'Mara, released a report entitled Megafires, which examined how climate change and other issues, including the funding issues at the United States Forest Service, are contributing to this growing problem.

I appreciate our witnesses being here to discuss the consequences of wildfires, air quality being chief amongst them, as well as some of the potential mitigation options such as more proactive forest management. But we do ourselves a disservice if we continue to hold hearings only looking at the effects of these fires while ignoring the underlying causes, including climate change that will continue to exacerbate this problem.

Thank you again, Mr. Chair, and I yield the remainder of my time to my good friend and colleague, Representative Matsui of California.

Ms. Matsui. Thank you, Ranking Member Tonko, for yielding. And I want to thank the witnesses for being here today.

I appreciate the subcommittee is holding a hearing on this important issue. California has had a historic year for fire. The Mendocino Complex fire consumed over 410,000 acres, burning for more than a month, and becoming the largest in our State's history. The Ferguson Fire took the lives of two brave firefighters and closed Yosemite National Park for 20 days. And the Carr Fire destroyed over 1,000 homes near Redding, north of my district.

While my district was fortunate and did not directly endure a wildfire this summer, Sacramento residents still had to contend with the smothering impacts of wildfire smoke. We had a record-breaking streak of 15 consecutive spare-the-air days when air quality was so poor that our air district encouraged people to stay inside and reduce pollution in any way possible.

If we don't take meaningful steps to reduce the risk and intensity of wildfires, then we will continue to face these overwhelming health, safety, and environmental challenges. That means we must adopt a sustainable approach to wildfire risk reduction. Management policies must recognize the impacts of climate change and the need to sustainably reduce the fuel load in our forests, ultimately moving their condition towards the pre-fire exclusion baseline.

Thank you, and I look forward to hearing the testimony from our witnesses.

I yield back.

Mr. Tonko. And I yield back our remaining 8 seconds. There you go.

Mr. Shimkus. The gentleman yields back his time.

The chairman is running over here. The ranking member, I can see, is still on the floor. So we will begin with our witnesses and then interrupt as we can.

We want to thank you all for being here today, taking the time to testify before the subcommittee. Today's witnesses will have the opportunity to give opening statements followed by a round of questions from members. Our witness panel – and I have already announced the panel. So I would like now to turn to Mr. Baertschiger, Oregon State Senator. And I am sure

Congressman Walden from Oregon will get here for most of your opening statement.

You are recognized for 5 minutes.

STATEMENT OF HERMAN BAERTSCHIGER JR., SENATOR, OREGON STATE SENATE; MARY ANDERSON, MOBILE AND AREA SOURCE PROGRAM MANAGER, AIR QUALITY DIVISION, IDAHO DEPARTMENT OF ENVIRONMENTAL QUALITY; SONYA GERMANN, STATE FORESTER, MONTANA DEPARTMENT OF NATURAL RESOURCES AND CONSERVATION, FORESTRY DIVISION; COLLIN O'MARA, PRESIDENT, NATIONAL WILDLIFE FEDERATION; AND TOM BOGGUS, STATE FORESTER, DIRECTOR OF TEXAS A&M FOREST SERVICE

Statement of Herman Baertschiger Jr.

Mr. Baertschiger. Thank you, Chairman Shimkus, Ranking Member Tonko, and members of the committee. Thank you for letting me have the opportunity to testify before you today about wildfires and their impact on my constituents and the people of Oregon.

The lingering effects of smoke and large fires impact thousands of people in my State every year. Immediate suppression of wildland fires during peak fire season would alleviate the impacts to our communities. In exchange for a suppression model, we must be conscious of the fact that our forests still need management, and fire is one of those management tools. But this can be accomplished outside of fire season by controlled burning. Smoke from controlled burns is far less impactful to my constituents than these large fires during the summer months.

Other management activities, including commodity production, logging, field reduction, are also effective in reducing the risk of severe fire.

My name is Herman Baertschiger, and I am an Oregon State Senator representing southern Oregon. My background is in forestry and wildland firefighting. In more than four decades of firefighting in the west, I have never seen a catastrophic high-intensity wildfire benefit our forests. However, I have seen many examples of low-intensity fire benefit our forests.

Fire has always been with us, and that is not going to change, likely. Large fires have affected the American people throughout our history. The fires of 1910 in Idaho, Montana, and Washington that burned 3 million acres changed how the U.S. Forest Service addressed fires. In Oregon, the Tillamook fires that occurred in the coast range four times between 1933 and 1951 forced Oregon also to address wildland fires. This approach is what, at times, having – is having us fighting large fires rather than suppressing small fires.

The aggressive fire suppression model changed about 30 years ago with the U.S. Forest Service. It changed from a fire suppression to a fire management. The comparison of fire suppression against fire management is best shown in a comparison of firefighting divisions of Oregon Department of Forestry and the U.S. Forest Service.

Oregon Department of Forestry has always employed an aggressive initial attack and suppression approach. The comparison of lands managed shows a shocking disparity between the two styles of firefighting. The U.S. Forest Service protects about 17 million acres of Oregon forestlands. And so far in 2018, 300,000 of those acres have burned. Oregon Department of Forestry protects about 16 million acres of forestlands in Oregon, and so far, only 70,000 of those acres have burned.

The two agencies protect about the same number of acres in Oregon but are having very different outcomes.

Also, the human factor can't be ignored. With over 300 million people in this country, we should expect more human-caused fire starts. Some people say that 9 out of 10 fires have a human element.

Due to severe wildfires, the lack of forest management and the different approach to firefighting, our communities have suffered weeks from toxic smoke. This year's citizens in southern Oregon endured 34 days of unhealthy air quality, and Travel Oregon estimated last year that $51 million was lost from smoke in tourism dollars. The Shakespeare festival in Ashland has lost over $2 million this year; Hell's Gate excursion, $1.5 million. Smoke has led to cancellation and delays of school activities, church activities, and other events.

To provide our citizens with relief from catastrophic wildfire, Congress should take action to promote active forest management and provide oversight and assure accountability over the U.S. Forest Service.

Managing fire during peak fire season to treat fuels is no longer acceptable. We cannot manage our forests during peak fire season with fire at the expense of the health and welfare and the economic viability of our communities. We have got to do something else.

I appreciate this opportunity to testify, and I welcome any questions that you may have.

[The prepared statement of Mr. Baertschiger follows:]

TESTIMONY OF OREGON STATE SENATOR HERMAN BAERTSCHIGER, JR. BEFORE THE HOUSE COMMITTEE ON ENERGY AND COMMERCE SUBCOMMITTEE ON THE ENVIRONMENT, SEPTEMBER 13, 2018

Chairman Shimkus, Chairman Walden, and Ranking Member Pallone, thank you for the opportunity to testify before you today about wildfires and their impact on the constituents of my district and all of Oregon. The lingering effects of smoke from large forest fires impact thousands of people every year. Congress Walden and I have toured the wildfires together and have seen and experienced the devastation these wildfires have on our communities. Immediate suppression of wildland fires during peak fire season would alleviate the impacts to our communities, their health and their economic viability. In exchange for a suppression model we must be conscious of the fact that our forests still need management and fire is one of those management tools. This can be accomplished outside of fire season by controlled burning, smoke from controlled burns is far less impactful to the communities than those of large fires during fire season.

My name is Herman Baertschiger, Jr. and I am an Oregon State Senator representing Southern Oregon. My background is in forestry and wildland firefighting. For 16 years I have been a wildland fire training instructor and I have been certified through the Oregon Department of Forestry as a National Type 3 Incident Commander and this year is my 41st fire season. In more than four decades of fighting fire from Washington to Montana, from California to Colorado, I have never seen a catastrophic wildfire benefit our forests.

Fire has always been with us and that is not likely to change, what has changed is the way we react to fire. Large fires have affected the American people for nearly two hundred years, in 1825, when the Great Miramichi Fire in New Brunswick burned over 3 million acres and killed 160 people. In 1871 the Peshtigo Fire in Wisconsin burned 1.2 million acres and killed 1,182 people. On the same day in 1871 in Illinois, Mrs. O'Leary's cow knocked over a lantern and started the Great Chicago Fire that killed hundreds and burnt half the City of Chicago. The fires of 1910 in Idaho, Montana and Washington that burned 3 million acres changed how the US Forest Service addressed fires. In Oregon the Tillamook forest fires that occurred four times between 1933 and 1951 forced Oregon to address wildfire also. Five of the last six years have seen enormous fires, in excess of the ten-year average. In 2018 we are fighting fires in the same area that we fought them in the previous year. To say that climate changes is the cause of catastrophic wildfires is incorrect. Our approach and management of the fires and the smoke they produce is what has changed in the last twenty years. This approach is what has us fighting large fires, rather than suppressing small fires.

After the Western fires of 1910, the US Forest service adopted a model of fire suppression. In the industry we have always referred to this with the term "out by 10," meaning that once a fire had started, it would receive whatever resources and attention needed to put it out by 10 AM the following morning. This aggressive fire suppression changed about twenty years ago to a fire management model. Today the US Forest Service often manages fire along with their suppression efforts.

The human factor can't be ignored when looking at the problem of wildland fires. In 1910 the US population was 92 million. In 2010, one hundred years later, it was 310 million. With a tripling of the population, we should expect more human caused fire starts. Recently the Oregon Department of Forestry estimated that the number of acres burned by human caused fires in the last ten years has doubled. With increases in fire starts, we should expect increases in fire severity, and increases in smoke effects from those fires.

The comparison of fire suppression against fire management is best shown in the comparison of the firefighting divisions of the Oregon Department of Forestry (ODF) and the US Forest Service (USFS). ODF has always employed an aggressive initial attack and suppression approach to wildfire.

The USFS had a similar approach until about the mid-1980's. Now thirty years later, the comparison of lands managed, and acres burned shows a shocking disparity between the two styles of firefighting. The USFS protects 17 million acres (+/-) of Oregon forest lands. In 2018, 300,000 acres of those protected forests were burned. ODF protects 16 million acres (+/-) of Oregon forest lands. In 2018, 70,000 acres of those protected forests were burned (these are estimates as the 2018 Fire Season is not complete).

These two agencies protect about the same number of acres of forest in Oregon but are having very different outcomes from their firefighting efforts. Another consideration is the impact on private landowners, in 2018 more than 33,000 acres of private land has burned by fires that started on federal lands.

Impacts of Wildfire and Smoke to Our Communities

Due to severe wildfires, the lack of active forest management and the U.S. Forest Service's current approach to firefighting, our communities have suffered from weeks of toxic wildfire smoke. As of Aug. 30, citizens in southern Oregon's Rogue Valley endured 24 days of "unhealthy" to "very unhealthy" air quality. This is the longest period of unhealthy air quality in the Rogue Valley since the Environmental Protection Agency began keeping air quality index records in 2000.

Travel Oregon estimated the state lost about $51 million in tourism revenue from wildfires and smoke last year, and this year's wildfire season will likely bring greater losses. Last month's wildfires forced airlines at the Rogue Valley Airport to cancel multiple flights and put delays on others. The smoke has created havoc and devastation for many businesses connected to Southern Oregon's vital Travel and Tourism Industry. The Oregon Shakespeare Festival in Ashland has already canceled more shows this year than it did in all of 2017. These are the first two years in a row that the festival has had to cancel multiple performances because of smoke. In 2018 the festival has lost over $2 million dollars, with over 20 performances canceled. Britt Festivals has moved its orchestra rehearsals and some performances from its hillside, Britt Pavilion amphitheater in Jacksonville, to the North Medford High School auditorium, which downsized their attendance from 2,200 attendees in auditorium to an 800-person capacity. The Rogue Valley Softball Association Fall Classic softball tournament was cancelled, marking the second consecutive year it's been called off. The wildfires and smoke have harmed our seasonal tourism businesses with many reporting 40% decreases in attendance, some as high as 80% lower attendance. Sections of the Rogue River have been closed at times, impacting those who are dependent on rafting and fishing. For July, Crater Lake was down on visitation by 22%, which represents over 50,000 people. Locally, the wildfire smoke led to cancellations or delays of school athletics, church activities and other events.

These smoky summers continue to tax our public health resources and health care system. In past years, local emergency departments reported an increase of patients suffering from extremely sore throats, headaches, burning eyes and significant respiratory distress.

Suggested Solutions

To provide our citizens relief from catastrophic wildfires and smoke, Congress should take action to promote active forest management and provide oversight and assure accountability over U.S. Forest Service wildfire management activities.

There is abundant science supporting the benefits of management activities-including logging, thinning and controlled burning-to improving the health and resiliency of our forests. According to the U.S. Forest Service's Fuels Treatment Effectiveness Database, 90 percent of fuels reduction projects were effective in reducing wildfire severity.[1] Researchers from the University of Montana found that comprehensive treatment prescriptions designed to restore sustainable ecological conditions can move 90 percent of treated acres into a low-hazard condition.[2]

Despite these benefits, federal agencies are failing to treat fire-prone forests at a pace and scale necessary to change the trend of larger and more severe fires. As much as 80 million acres of National Forests System lands are at a high, to very high, risk of catastrophic wildfire. And the Forest Service is only treating between 1 and 2 percent of high risk acres per year. That's why congressional action is needed to address the *three primary barriers* to active forest management on federal lands:

The first barrier to active forest management is the lack of funding to prepare forest projects and timber sales. The U.S. Forest Service often lacks the funding and personnel to develop and implement projects that reduce the

---

[1] USFS, Adaptive Management Services Enterprise Team, Fuels Treatment Effectiveness Database (fs.fed.us/adaptivemanagement).
[2] C. Keegan, C. Fiedler, T. Morgan. Wildfire in Montana: Potential hazard reduction and economic effects of a strategic treatment program, Forest Products Journal, July/August 2004).

risks of wildfires, insects and disease. Today more than half of the agency's budget is consumed by escalating wildfire suppression costs, which itself is due to the lack of management on our overgrown and fire-prone federal forests. The U.S. House of Representatives addressed this barrier by approving the Consolidated Appropriations Act of 2018 ("Omnibus"), which provides a new disaster cap allocation for wildfire suppression beginning in 2020. This will end the practice of "fire borrowing" that has forced the Forest Service to temporarily redirect funds away from preventative forest health programs. I also applaud the U.S. House for increasing funding for the Forest Service's Hazardous Fuels line items in the omnibus spending bill, and for increasing funding for the agency's timber sale program in the House Interior Appropriations package for FY 2019. If enacted, the House level would represent a 20 percent increase in the timber program since FY 2013. The House Report urges the Forest Service to offer a 4 Billion Board Foot sale program, a level that hasn't been reached in a quarter century.

The second barrier to active forest management is the significant cost and time it takes for federal agencies to satisfy environmental analysis and compliance requirements. At a time when the Forest Service struggles to fund wildfire suppression activities, the agency spends more than $356 million annually just to satisfy National Environmental Policy Act (NEPA) analysis and compliance requirements on forest management projects.[3] It can take Forest Service personnel 18 months to four years to satisfy these requirements for a single forest project, and as a result they spend 40 percent of their time doing paperwork instead of managing forests. The

U.S. House addressed this barrier by approving the Resilient Federal Forests Act (H.R. 2936), and by approving strong provisions in the Agriculture and Nutrition Act of 2018 (H.R. 2), known as the Farm Bill. Both measures provide our land management agencies with expanded categorical exclusions under NEPA to expedite treatments on forests that are

---

[3] Feasibility Study of Activities Related to National Environmental Policy Act (NEPA) Compliance in the US Forest Service-Final Report, USDA Forest Service Competitive Sourcing Program Office Washington, DC. Available at https://www.peer.org/assets/docs/fs/08_14_1_nepa_feasibility_study.pdf.

at immediate risk of wildfire, insects and disease, and to protect municipal watersheds. Providing these NEPA efficiencies will help reduce the cost and time required to plan forest projects, and will provide and direct more resources toward improving the health of our forests.

The third factor is the obstruction and litigation that typically stalls much of the work that needs to be done on our federal lands. Our federal land agencies are often paralyzed by the real and perceived threats of litigation and process obstruction. In U.S. Forest Service Region 1, for example, it is estimated litigation has encumbered up to 50 percent of planned timber harvest volume and treatment acres. Once again, the U.S. House addressed this barrier by approving H.R. 2936 that prohibits litigant groups from receiving attorney fees when they sue to stop a project that is intended to reduce the threats of wildfire and insect infestations. It also requires that any court hearing a case regarding a Forest Service action must weigh the benefits of taking short-term action versus the potential long-term harm of inaction, such as the threat of catastrophic wildfire. As an alternative to costly litigation, the legislation also proposes an innovative pilot project to test the use of arbitration to address challenges to certain forest management activities.

Until environmental litigation is addressed, American taxpayers will continue to carry the increased burden of higher firefighting and land management costs, toxic smoke, and the continued loss of forests and the benefits they provide.

Managing wildland fire during peak fire season to treat fuels as a way to manage our forests is no longer acceptable. We cannot manage our forests with fire at the expense of the welfare and economic viability of our communities. We have to do something else.

Once again I appreciate the opportunity testify and welcome any questions you may have.

Mr. Shimkus. The gentleman yields back his name.

The chair now recognizes the chairman of the full committee, another Oregonian, Chairman Walden, for 5 minutes.

OPENING STATEMENT OF HON. GREG WALDEN, A REPRESENTATIVE IN CONGRESS FROM THE STATE OF OREGON

The Chairman. Thank you very much, Mr. Chairman. I apologize for being a little late. We had the WRDA bill on the floor where a number of our provisions, including Safe Drinking Water Act and some provisions for drought relief in the Klamath Basin were before the House, so I needed to speak on that before coming here.

I want to thank you for holding this hearing, and I want to thank our witnesses for being here.

Today's hearing focuses on this topic that you have already heard from the Senator about, of great concern to Oregonians and those across the west who are experiencing terrible air quality. Hazardous, dangerous, unhealthy air quality smoke from these wildfires is literally choking people to death.

In my home State of Oregon alone, we have already seen over 700,000 acres destroyed by fire. These fires have left communities in my district blanketed with smoke and with the worst air quality in the world, period. Stop. Medford, Oregon, experienced the worst run of unhealthy air quality since the EPA began making such determinations in 2000.

The leading offender is particulate matter. An article in the New England Journal of Medicine in March pointed out the robust evidence linking exposure to particulate matter to cardiopulmonary mortality and issues with asthma and COPD. I heard from a woman yesterday on a tele-town hall: COPD. She was just getting out of the hospital all as a result of this smoke.

According to EPA research, premature deaths tied to wildfire air pollution were as high as 2,500 per year between 2008 and 2012. Other research out of Colorado State University suggests it could be as high as 25,000 people per year die prematurely because of this smoke. This is a life-and-death matter in the west.

Making matters worse, it is hard to escape the smoke even in your own home. Curt in Eagle Point dropped off his air filter from his CPAP machine. I have got a picture of it up there. That filter is supposed to last for 2 weeks.

That is, I believe, 2 days. You can see it up there and how dirty it got within 2 days inside his home during these fires.

Or take this car cabin air filter. It was replaced after 2 months during the fire season. You can see up on the screen what a new one looks like. Two months, that is what it looked like in his car.

Nearly three decades of poor management have left our Federal forests overstocked with trees and vegetation that fuel increasingly intense fires. Stepping up active forest management practices such as thinning, prescribed fire, and timber harvest, one of the best ways we could reduce the fuel loads and, therefore, the impact of the smoke from wildfires.

Sadly, bureaucratic red tape, obstructionist litigation by special interest groups, it has all added up to make it very difficult to implement these science-based management techniques that we know work. And we are left to choke on the resulting wildfire smoke.

In 2017, the number of fires started on lands protected by the Oregon Department of Forestry and the U.S. Forest Service Land were split nearly 50/50. Forest managed lands, however, accounted for over 90 percent of the acres burned. So that is the Federal ground. This is partly due to forest management but also how fires are fought.

As fires are managed rather than suppressed and back burned acreages increased, there is a clear impact on air quality and, therefore, on the air quality and health of our citizens. These agencies need to do more to take this into account when they make their decisions.

As devastating as it is in the summer months, fire can also be a management tool. We know that. Prescribed fire, after mechanical thinning, can help reduce fuel loads and reduce emissions by up to 75 percent, if it is done at the right time and the right way. State smoke management plans set the process for these burns with an aim to protect public health, but also limit the work that gets done. According to Forest Service data, smoke management issues limited between 10 and 20 percent of their prescribed fire projects last year in Oregon.

I look forward to hearing from our witnesses today about your perspectives on these issues and how we get the right balance. I also want to thank Senator Baertschiger for joining us from Oregon. He is the co-chair of the bipartisan fire caucus in Oregon, has nearly 40 years of experience in wildland fire and forest management both. So thanks for flying out to be here. And just to conclude, I would like to share a message I received from Jennifer. She is a mother in Medford, Oregon. Jennifer said: As a native Oregonian, living in southern Oregon my entire life, I write to express my extreme frustration with Oregon's lack of forest management. This is now the third or fourth year that we are hostages in our own homes, that my children are robbed of being able to play outside. I absolutely hate that nothing is done to prevent this from happening.

Well, we are here to help the concerns I hear from people like Jennifer and families across my district who have one simple message: Something needs to change.

And in conclusion, I just got an email from a friend of mine in Medford, who is on the Shakespeare board, the Oregon Shakespearian Theater board in Ashland. And they said: I have exciting news. Our safety, health, and wellness manager sent this update. We are officially closing the smoke watch that started back on July 18 and returning to normal operations.

I believe they had to cancel 25 outdoor plays at the Allen Elizabethan, and one for the Bowmer, for a total of 26 cancellations for performances. And so this is a real bad thing for the economy. It is bad for our health.

Mr. Chairman, I appreciate your holding this hearing, and I thank the witnesses for being here. And I yield back.

[The prepared statement of Chairman Walden follows:]

OPENING STATEMENT OF CHAIRMAN GREG WALDEN SUBCOMMITTEE ON ENVIRONMENT HEARING, "AIR QUALITY IMPACTS OF WILDFIRES: MITIGATION AND MANAGEMENT STRATEGIES" SEPTEMBER 13, 2018

Today's hearing focuses on a topic of great concern to Oregonians and those across the West who are experiencing terrible air quality and choking on smoke from wildfires. In my home state of Oregon alone, we've already seen over 700 thousand acres destroyed and the fires are still burning.

These fires have left communities in my district blanketed with smoke and with the worst air quality in the world. Medford, Oregon experienced the worst run of "unhealthy" air quality since EPA began recording in 2000.

A leading offender is particulate matter. An article in the New England Journal of Medicine in March pointed out the robust evidence linking exposure to particulate matter to cardiopulmonary mortality and issues with asthma and COPD. According to EPA research, premature deaths tied to wildfire air pollution were as high as 2,500 per year between 2008 and 2012. Other research out of Colorado State University suggest it could be as high as 25,000 people a year.

Making matters worse, it is hard to escape the smoke, even in your own home. Curt in Eagle Point, OR dropped off this air filter from his C-PAP machine. He had to replace it after two days – it is supposed to last two weeks.

Or take this car cabin air filter that was replaced after two months during fire season. A new one looks like this. You begin to realize what people are suffering through.

Nearly three decades of poor management has left our federal forests overstocked with trees and vegetation that fuel increasingly intense fire seasons. Stepping up active forest management practices, such as thinning, prescribed fire, and timber harvests, is one of the best ways to reduce the fuel and the impact of smoke from wildfires.

Sadly, bureaucratic red tape, and obstructionist litigation by special interest groups, has made it difficult to implement these science-based management techniques. And we're left to choke on the resulting wildfire smoke.

In 2017, the number of fire starts on lands protected by the Oregon Department of Forestry and those on U.S. Forest Service land were split nearly 50/50. The Forest managed lands, however, accounted for over 90 percent of the acres burned. This is partly due to forest management, but also how fires are fought.

As fires are managed, rather than suppressed, and back burned acreages increase, there is a clear impact on air quality and our health. These agencies should do more to take that into account.

As devastating as it is in the summer months, fire can also be a management tool. Prescribed fire after mechanical thinning, can help reduce fuel loads and reduce emissions by up to 75 percent. State Smoke Management Plans set the process for these burns with an aim to protect public health, but also limit the work that gets done. According to Forest Service data, smoke management issues limited between 10 and 20 percent of their prescribed fire projects last year in Oregon. I look forward to hearing from our witnesses today on whether these plans properly balance the risk from prescribed fire with the risk of far more intense wildfire.

I also want to thank Senator Herman Baertschiger for joining us from Oregon. Senator Baertschiger is co-chair of the bipartisan fire caucus in Oregon, and has nearly 40 years of experience in wildland fire and forest management. Thank you for your participation and sharing your knowledge with us today.

Just to conclude, I'd like to share a message I received from Jennifer, a mother in Medford. Jennifer said, "As a native Oregonian living in Southern Oregon my entire life I am writing to express my extreme frustration with Oregon's lack of forest management. This is now the third or fourth year that we are hostages in our own homes, that my children are robbed of being able to play outside. I absolutely hate that nothing is done to prevent this from happening."

We are here today to help address the concerns I hear from people like Jennifer and families across my district who have one simple message: something needs to change.

***

Mr. Shimkus. The gentleman yields back his time.

The chair now would like to recognize Ms. Mary Anderson, mobile and area source program manager, Air Quality Division, Idaho Department of Environmental Quality.

You are recognized for 5 minutes.

STATEMENT OF MARY ANDERSON

Ms. Anderson. Thank you for the opportunity to provide some insight into how wildfires are impacting Idaho citizens.

Wildfires are the single largest air pollution source in Idaho. In the past, Idaho would experience severe wildfire season with heavy localized air quality impacts every 3 to 4 years. Now, we are seeing heavy regional air quality impacts every year from large, sometimes catastrophic wildfires in Idaho, central and northern California, Oregon, Washington, Nevada, and British Columbia. These catastrophic wildfires caused by fuels that have cumulated as a result of active fire suppression, drought, and climate change.

In 2017, wildfire smoke caused widespread impacts starting in early August. And by the first week of September, smoke thoroughly blanketed all of Idaho, exposing many Idaho citizens to potentially serious health impacts.

About 700,000 acres were burned in Idaho in 2017. Idaho is also surrounded by wildfires, meaning wind from any direction brought smoke into the State. Nearly 5.5 million acres burned in neighboring States and British Columbia in 2017. All these fires had direct impacts on Idaho residents at one time or another throughout the wildfire season. We are seeing similar impacts this year.

What I have described above is now the new norm. The public now experiences smoke impacts throughout the summer every year, with periods of very unhealthy to hazardous air quality conditions. To deal with the smoke impacts, the public wants information so they can make decisions to protect themselves and, in the case of schools, those they are responsible for. Telling them to remain indoors and limit exposure is no longer sufficient. In many cases, the air quality indoors is just as bad or worse than the air quality outside.

Responding to wildfire smoke impacts requires significant resources from DEQ and other agencies throughout Idaho. To properly respond to wildfires and mitigate health impacts from smoke, the communities that are repeatedly hard hit from wildfire smoke must be made smoke ready before the smoke event occurs. This means working with communities to identify tools citizens can use to protect themselves from the smoke.

An example of a smoke ready community action is identifying the sensitive population, such as elderly people with lung or heart issues, and purchasing a cache of room-sized HEPA filters prior to the wildfire season so they can be distributed at the start of the emergency. Establishing a smoke ready community must be done prior to the wildfire season in order to respond to the emergency in a timely manner.

To be effective, smoke ready communities require funding similar in the way – similar to the way firewise programs are funded. Funding for both these programs would allow communities to prepare for wildfires from both the fire safety and public health aspect.

We agree that prescribed fire is an important tool in reducing fuels that contribute to catastrophic wildfire, but prescribed fire also causes smoke that needs to be managed. When prescribed fire is being discussed as a way to mitigate wildfire impacts, it is important to remember that reasonable and effective smoke management principles and decisions must be used to truly lessen smoke impacts and not simply move smoke from one time of the year to another.

To manage smoke impacts from prescribed burning, the Montana/Idaho Airshed Group was created. This group implements a smoke management program for organizations that conduct large-scale prescribed burning and the agencies that regulate this burning.

Burn decisions in Idaho are very much driven and limited by the weather. Northern Idaho is very mountainous. Smoke from prescribed burning can sink into the valleys and impact communities. Using best smoke management practices requires good weather that will allow the smoke to rise up high into the atmosphere and disperse so as not to impact the public. The key to this airshed group is coordinating burn requests and approvals to looking at the regional picture, not just burns on an individual basis.

The Airshed Group uses a meteorologist to provide a weather forecast specific to prescribed burning. A coordinator evaluates all burns that are proposed, other burning, and emission sources occurring in the area, current and forecasted air quality, to determine if and how much burning can be approved. This process helps to ensure that smoke does not accumulate in valleys and impact the public.

DEQ works closely with the airshed group during the active burn season. We review the weather forecast, air quality data, and proposed burns, and provide recommendations to the airshed group on a daily basis.

There is no short-term quick fix. We need to address all causes of wildfire and look at new innovative solutions and mitigation strategies to address the matter. The key to success will be working in partnership with all stakeholders, air quality agencies, State and Federal land managers, large and small private prescribed burners, the general public, environmental groups, and others who use burning as a tool. The only way to make progress is to have an open, honest, and trusting dialogue based on facts and science.

Thank you.

[The prepared statement of Ms. Anderson follows:]

TESTIMONY OF MARY ANDERSON, MOBILE AND AREA SOURCE PROGRAM MANAGER, AIR QUALITY DIVISION IDAHO DEPARTMENT OF ENVIRONMENTAL QUALITY, BEFORE THE HOUSE COMMITTEE ON ENERGY AND COMMERCE SUBCOMMITTEE ON ENVIRONMENT, REGARDING AIR QUALITY IMPACTS OF WILDFIRES: MITIGATION AND MANAGEMENT STRATEGIES, SEPTEMBER 13, 2018

Good afternoon, Chairman Shimkus, Ranking Member Tonko, and members of the subcommittee. I am Mary Anderson, Air Quality Division program manager at the Idaho Department of Environmental Quality (DEQ). Thank you for the opportunity to testify today and provide some insight into how wildfires are impacting Idaho citizens.

Air Quality Impacts of Wildfires

Wildfires are the single largest air pollution source in Idaho and the entire Pacific Northwest. Fine particulate matter (PM2.5) levels are decreasing nationally but increasing in the Pacific Northwest and Northern Rockies due to wildfire smoke. Since the mid-1980s, the total US area burned by wildfires has been increasing, with fires in the Pacific Northwest accounting for over 50% of the increased acreage. The length of the fire season has grown along with the number, size, and duration of wildfires. This situation is forecasted to stay the same or worsen in the future. Numerous catastrophic wildfires have become the norm during our summer months, causing heavy regional air pollution events. These catastrophic wildfires are caused by fuels that have accumulated as a result of a century of active fire suppression, drought, and climate change.

In the past, Idaho would experience a severe wildfire season with heavy localized air quality impacts every 3-4 years, with low air quality impacts in the intervening years. Now we are seeing heavy regional air quality impacts every year from large, sometimes catastrophic wildfires in Idaho, central to northern California, Oregon, Washington, Nevada, and British Columbia.

In 2017, wildfire smoke caused widespread impacts in early August, with air quality reaching the Unhealthy for Sensitive Groups category in nearly every area of Idaho. It is now fairly common to see widespread impacts in August.

By the first week of September 2017, wildfire smoke thoroughly blanketed all of Idaho, exposing many Idaho citizens to potentially serious health impacts. The most severe air quality impacts were in northern and central Idaho, where Hazardous conditions were measured for four consecutive days in the Coeur d'Alene area of northern Idaho.

About 700,000 acres were burned by wildfire in Idaho in 2017. Idaho was also surrounded by wildfires, meaning wind from any direction brought smoke into the state. Nearly 5.5 million acres burned in neighboring states and British Columbia in 2017. All these fires had direct impacts on Idaho residents at one time or another throughout the wildfire season. Idaho wildfires alone released an estimated 111,000 tons of direct fine particulate pollution into the air (about 25 times the amount of fine particulate pollution emitted by all the cars and trucks in Idaho in a year).

Preliminary information for the 2018 wildfire season indicates that it is as bad as or worse than 2017. Idaho became heavily impacted by smoke around July 10 and experienced smoke impacts on a daily basis until the first full week of September. The most heavily impacted regions have been northern Idaho, central Idaho, southwest Idaho—including the densely populated Treasure Valley—and at times the rest of southern Idaho. This year, the majority of smoke came from fires outside of Idaho: Washington, Oregon, California, and southern British Columbia. The predominant weather patterns have allowed for consistent smoke brought into Idaho. We are seeing similar levels of air quality impacts as we did in 2017.

What I've described above is now the new normal. The public now experiences smoke impacts throughout the summer every year, with periods of Very Unhealthy to Hazardous air quality conditions. To deal with the smoke impacts, the public wants information so they can make decisions to protect themselves and, in the case of schools, those they are responsible for. The public wants information about how bad the air quality is, how long the smoke will last, and what precautions they should take. Telling them to remain indoors and limit exposure is no longer sufficient. In many cases, the air quality indoors is just as bad or worse than the air quality outside.

Local governments and school districts have had to develop policies to respond to the wildfire smoke impacts brought by catastrophic wildfires. During the 2012 wildfire season, very few if any school districts had air quality policies that determined outside activity based on current or forecasted air quality. In 2018, the majority of school districts have these policies and are making daily decisions on whether to have outside recess, hold practices indoors, and cancel football games.

To ensure a coordinated response to wildfire smoke, Idaho developed a Wildfire Smoke Response Protocol, similar to Oregon and Washington. This protocol identifies organizations, partners, and other governmental entities (city and county) that play important roles in the overall response to these wildfire smoke events. The protocol highlights general duties and responsibilities, provides examples of agency actions and assistance needed, and recommends public health actions based on level and duration of smoke exposure caused by wildfire smoke. Key participants are federal land managers (US Forest Service), the Environmental Protection Agency (EPA), tribes, DEQ, the Idaho Department of Health and Welfare, and public health districts. We also work closely with county emergency managers, Red Cross, and school districts to ensure a consistent message is communicated.

To meet the public's need for information, DEQ and cooperating agencies use many tools:

- Idaho Smoke Blog
- Social media (Twitter, Facebook, NextDoor)
- Websites
- News releases
- Videos explaining our smoke forecasts
- Daily smoke and air quality forecast email blasts

DEQ responds to local community requests and federal land manager requests to supply additional air quality monitoring in areas heavily impacted by wildfire smoke. Each year, DEQ deploys between one and four additional monitors throughout the state. This deployment requires significant coordination to identify and establish a suitable location for the monitor to operate. We have deployed four monitors so far in 2018.

Responding to wildfire smoke impacts requires significant resources from DEQ and other agencies throughout Idaho. To properly respond to wildfires and help mitigate health impacts from smoke, the communities that are repeatedly hard hit by wildfire smoke must be made "smoke ready" before the smoke event occurs. This means working with the communities and counties to identify tools citizens can use to protect themselves from the smoke. An example of smoke ready community action is identifying the sensitive populations (elderly, people with lung or heart issues) and purchasing a cache of room-sized HEPA filters prior to the wildfire season. Establishing a smoke ready community must be done prior to the wildfire season in order to respond to the emergency in a timely manner. To be effective, smoke ready communities require funding, similar to the way Firewise programs are funded. Funding for both these programs would allow the communities to prepare for wildfires from both the fire safety and public health aspect.

Many types of open burning occur in Idaho throughout the year. Prescribed burning—which includes burning for forest health, slash burning after a timber harvest, and rangeland burning—typically occurs in the spring and fall, outside the wildfire season. Prescribed burning is conducted by state and federal land managers and landowners. Agricultural burning also occurs during these times as does residential backyard burning. During the winter, a high percentage of the population in rural Idaho uses wood to heat their homes, which creates additional smoke impacts.

Prescribed fire is an important tool in reducing fuels that contribute to catastrophic wildfire, but prescribed fire also causes smoke that needs to be managed. When prescribed fire is being discussed as a way to mitigate wildfire impacts, it is important to remember that reasonable and effective smoke management principles and decisions must be used to truly lessen smoke impacts and not simply move smoke from one time of year to another.

Smoke from prescribed fire has the potential to jeopardize attainment demonstration in some of our nonattainment areas (areas not meeting air quality standards) if not applied appropriately. Data can be flagged as "exceptional," thereby excluding it from attainment demonstrations, but only if adequate smoke management principles are adopted and applied.

To manage prescribed burning smoke impacts, the Montana/Idaho Airshed Group was created to implement a prescribed fire smoke management program for organizations that conduct large-scale burning and the agencies that regulate this burning. The airshed group is an effective collaboration of state and federal agencies and most large private landowners to limit smoke impacts from prescribed fire while providing as much flexibility as possible. The airshed group is composed of three units: Montana, North Idaho, and South Idaho, formed in 1978, 1990, and 1999, respectively.

As a group, the members sign a memorandum of understanding and commit to abide by the group's operating guide, which details policies and procedures. The group's members are committed to working together to manage air quality in a responsible manner so as not to impact public health.

The key to the airshed group is coordinating burn requests and approvals to look at the regional picture, not just burns on an individual basis. The airshed group uses a meteorologist to provide a weather forecast specifically for prescribed burning. A coordinator evaluates all burns that are proposed, other burning/emissions sources occurring in the area, and current and forecasted air quality to determine if and how much burning can be approved. This process helps to ensure that smoke does not accumulate in valleys and impact the public.

DEQ works closely with the airshed group during the active burn season. We review the weather forecast, air quality data, and proposed burns and provide recommendations to the airshed group on a daily basis.

Currently, the airshed group is staffed with two meteorologists from federal agencies who provide part of their time to the airshed group. One US Forest Service staff person reviews all proposed burns in Montana and Idaho, coordinates with both Idaho and Montana DEQs, and makes a burn recommendation for each proposed burn. Staff are barely able to keep up with the workload at the current rate. If prescribed burning is increased, one person will not be sufficient to manage the workload.

Burn decisions in Idaho are very much driven, and limited, by the weather. Northern Idaho is very mountainous. Smoke from prescribed burning can sink into the valleys and impact those populations. Burn managers require specific weather conditions to ensure the burn accomplishes the intended goals. Using best smoke management practices also requires good weather that will allow the smoke to rise high into the atmosphere and disperse so as not to impact the public.

Burning, whether wildfire or prescribed, is a large part of Idaho's air quality concerns. In order to respond to, and effectively manage, all the planned open burning, DEQ is developing a comprehensive smoke management program that addresses all types of burning in a consistent manner statewide. This smoke management program has two goals. The first is to protect public health, which is a key component of DEQ's mission. We protect public health by ensuring open burning does not cause an exceedance of a National Ambient Air Quality Standard. The second goal is to provide flexibility to burners when it will not jeopardize public health.

Idaho also participates in a smoke management group facilitated by EPA Region 10. This group consists of air quality regulatory agencies; health agencies; private, state, and federal land managers; researchers; and interested stakeholders and strives to improve all aspects of wildfire response and smoke management of prescribed burning and agricultural burning.

According to a study funded by the Joint Fire Science Program, funding is one of the biggest hurdles for prescribed burning in Idaho. If funding is resolved, air quality could become the main hindrance. Air quality smoke management programs, and staffing, will need to adapt now to be ready to handle increased use if other issues are resolved.

Conclusion

There is no short-term, quick fix. We need to address all causes of wildfires and look at new, innovative solutions and mitigation strategies to address the matter. The key to success will be working in partnership with all stakeholders: state and federal land managers, large and small prescribed burners, the general public, environmental groups, and others who use burning as a tool. The only way to make progress is to have an open, honest, and trusting dialogue based on facts and science.

***

Mr. Shimkus. And the chair thanks the gentlelady.

The chair will now recognize the ranking member of the full committee, Congressman Pallone from New Jersey, for 5 minutes.

OPENING STATEMENT OF FRANK PALLONE, JR.

Mr. Pallone. Thank you, Mr. Chairman, and thank you for letting me – I know – I was on the floor with our chairman.

It has been a year now since this subcommittee last held a hearing on wildfires. And since that time, the same regions of the country are suffering due to the large number and size of forest fires, causing tremendous damage. And this is, once again, particularly destructive to Western States.

We have all seen the devastating images of lives lost and homes destroyed. These extreme wildfires are also creating poor air quality in States far away from the fires.

Last month, the National Weather Service found that smoke from western wildfires has spread as far as New England. And these wildfires are tragic, but they should not be a surprise. For years, scientists have warned that climate change was very likely going to contribute to the increased fire intensity and frequency that we are seeing now. That is exactly what we are seeing, and we are not going to improve the situation by only looking at forest management or timber harvesting practices.

If this Congress wants to truly address the increase in extreme wildfires, we must act to slow the global warming that is driving changes in climate and weather patterns.

Unfortunately, the Trump administration and congressional Republicans refuse to address climate change and have instead pushed policies that will exacerbate our climate problems. Here is my list of President Trump's most significant climate actions.

First, he pulled the U.S. out of the Paris Agreement, giving up our spot as a global leader and turning his back on our allies. Then he proposed to replace the commonsense Clean Power Plan with a dirty power scam that lets polluters off the hook.

The EPA even admits this proposal will result in 1,400 more premature deaths every year. Third, President Trump proposed to relax standards for fuel efficiency in vehicles, hurting consumers and ensuring more climate changing substances are emitted into the air. And fourth, he doubled down on a loophole in the Clean Air Act that allows more efficient and polluting heavy duty trucks on our roadways.

And then just this week, Trump relaxed controls on methane pollution from oil and gas operations and landfills. The President has also blocked all Federal agencies from considering or acknowledging the costs associated with climate change when making decisions, and he has proposed to cut funds for energy efficient programs and support for renewable energy. And finally, he continues to threaten to abuse emergency authorities to subsidize the oldest and least efficient coal plants in the country.

President Trump and his administration are doing everything possible to increase emissions and block any attempt to slow the rate of climate change. The result is rising seas, extreme weather events, severe drought and, of course, extended and intense fire seasons. And these are costing lives, destroying property and infrastructure, and costing us billions in disaster assistance.

And as we sit here, the southeast is about to be hit by another powerful hurricane devastating more communities. A new report from the researchers of Stony Brook University and Lawrence Berkeley National Laboratory finds that hurricane Florence is about 50 miles wider as a result of climate change. That means that hurricane can result in 50 percent more rainfall. Yet Republicans refuse to address climate change.

Even here today, the focus is not where it should be. How many more of these events do we need before Republicans join us in taking decisive action to combat climate change? When are Republicans going to stop actively pursuing policies that make the problem worse?

If we are serious about stemming the terrible growth of the forest fire season as well as these other natural disasters, we need to abandon the disaster that is the Trump administration climate policy, and we need to do it immediately.

And with that, Mr. Chairman, I yield back. And thank you for the time.

Mr. Shimkus. The gentleman yields back his time.

The chair now recognizes Ms. Sonya Germann, State forester, Montana Department of Natural Resources and Conservation, Forestry Division, on behalf of the National Association of State Foresters.

You are welcome and recognized for 5 minutes.

STATEMENT OF SONYA GERMANN

Ms. Germann. Thank you, Chairman Shimkus, Ranking Member Tonko, Full Committee Chair Walden, Ranking Member Pallone, and members of the subcommittee. It is a true honor to be before one of the Nation's longest standing committees to discuss wildfire impacts to air quality and strategies we are undertaking to mitigate those impacts.

My name is Sonya Germann, State forester of Montana. And like Mr. Boggus, I am here testifying on behalf of the National Association of State Foresters. I am also a member of the Council of Western State Foresters, which represents 17 Western States and six U.S.-affiliated Pacific islands. I have spent my life in Montana in the past 12 years in forestry, with an emphasis on active forest management, and I am honored to share the Montana perspective with you here today.

The 2018 fire year has been challenging, not only in severity and duration, but most importantly in the number of lives lost. There have been 14 fire-related fatalities, a devastating loss to families, the wildland firefighting community, and the greater public. Across the Nation and particularly in the west, wildfires are growing more intense and so large we are now calling them megafires.

In Montana, our fire season is, on average, 40 days longer than it was 30 years ago. And as the chairman suggested, more than 7 million acres has burned since January 1 on a national scale. And let me put that in perspective for you.

In the past 16 years, we have surpassed the 7 million acre mark eight times and the 9 million acre mark five times. In the years prior to that, we reached 7 million acres only once.

Although the 2018 fire year in Montana has thankfully been relatively moderate, our citizens and wildland firefighters are still reeling from 2017, which was our most severe season on record since 1910 with over 1.2 million acres burned, which is an area roughly the size of Delaware.

With severe fire years comes intense smoke. And according to the Montana Department of Environmental Quality, the air quality standards for particulate matter have been exceeded 579 times for wildfire over the past 11 years, with 214 of those occurring in 2017.

Fire is a natural part of our ecosystem. What is not natural are the unprecedented forest conditions we are facing. Nearly a century of fire exclusion has led to excessive fuel loading and changed forest types. These factors, in addition to insect epidemics, persistent drought, and climate change have resulted in a disproportionate amount of Montana's fire-adapted forests being at significant risk of wildfire. Today, over 85 percent of Montana's forests are elevated wildfire hazard potential.

As land managers, we understand the connection between fuels, wildfires, severity, and smoke. Consequently, we make concerted efforts to work with key partners to reduce fuels that in turn reduce wildfire risk and smoke-related impacts. Treatments like prescribed fire mechanical fuels reduction will not prevent wildfires from occurring but can influence how a wildfire burns. Experience shows that actively managed forested stands often burn with less intensity and produce less smoke than stands with higher fuel loading. Additionally, active fuels reduction can create safer conditions for wildland firefighters and may also offer crews opportunities to keep those fires smaller.

Along with our key partners, we endeavor to get more prescribed fire and mechanical fuels reduction work done on the ground. And as Ms. Anderson described, we are a part of the Montana/Idaho Airshed Group. This group assures coordinated compliance with regulatory agencies and strives to help us accomplish more prescribed burning while complying with air quality standards.

In Montana, proof is in the air quality data. Over the past years, prescribed fire has exceeded air quality standards only four times compared to 579 for wildfire. This group has been recommended as a model for other States to follow.

And lastly, with over 60 percent of forested land in Montana managed by Federal agencies, we strongly support authorities that facilitate fuels reduction projects and allow them to be completed more quickly through collaborative action. The Good Neighbor Authority and categorical exclusions for wildfire resilient projects represent two such authorities.

We strongly appreciate and value Congress' efforts to make authorities like these available to our Federal partners.

In closing, my written testimony has been made available to you, and I look forward to answering any questions you may have. Again, thank you for the opportunity to testify before you today.

[The prepared statement of Ms. Germann follows:]

TESTIMONY OF SONYA GERMANN, MONTANA STATE FORESTER, ON BEHALF OF THE NATIONAL ASSOCIATION OF STATE FORESTERS, SUBMITTED TO THE U.S. HOUSE COMMITTEE ON ENERGY AND COMMERCE, SUBCOMMITTEE ON THE ENVIRONMENT, FOR HEARING ON AIR QUALITY IMPACTS OF WILDFIRES: MITIGATION AND MANAGEMENT STRATEGIES, SEPTEMBER 13, 2018

Good morning, Chairman Shimkus, Ranking Member Tonko, and Members of the subcommittee. My name is Sonya Germann. I am the State Forester for the State of Montana, Department of Natural Resources and Conservation, Forestry Division.

I appreciate the opportunity to speak with you today and submit written testimony as the Committee considers the significant impacts of wildfire smoke on citizens and communities across the country, as well as the preventive role prescribed fire and hazardous fuels reduction can have in mitigating smoke impacts from unplanned wildfires.

The National Association of State Foresters (NASF) represents the directors of the state forestry agencies in all 50 states, eight territories, and the District of Columbia. State forestry agencies, like mine, deliver technical and financial assistance; protect lives, property and natural resources from wildfire; and conserve forest health, water, and other ecosystem services on more than two-thirds of our nation's 766 million acres of forests. Through the State Fire Assistance (SFA) and Volunteer Fire Assistance (VFA) programs, state agencies equip prescribed fire managers and wildfire response resources who work on state and private lands, where over 80% of the nation's wildfires start.

In addition, state forestry agencies work closely with our federal partners in managing complex multi-jurisdictional landscapes. For example, with the authority granted by Congress in the 2014 Farm Bill, over 30 states, including in the state of Montana, have signed Good Neighbor Authority (GNA) agreements with the federal government. Under GNA, states may use their own contracting procedures to serve as the agent of the U.S. Forest Service (Forest Service) to conduct restoration activities on federal lands and adjacent non-federal lands. By partnering together in this way, the Forest Service and states like Montana can better leverage resources to accomplish more restoration work on the ground.

While the duties of state agencies vary, each share common forest management and restoration missions and most have statutory responsibilities to provide wildland fire protection on public and private lands. As such, we are intimately aware of the increasing occurrence of wildland fire and associated smoke impacts across the country.

SUMMARY OF NATIONAL AND REGIONAL FIRE ACTIVITY

According to the National Interagency Fire Center, the 2018 fire year has proven challenging both in its severity and its duration. However, the severity and duration of the 2018 fire year is more consistent with the past decade than it is an anomaly. Since 2002, with very few exceptions, fire seasons have tended to be more active, with larger acreages burned and more severe conditions than any other decade since we began keeping consistent and accurate records in 1960.

Quantifying a fire season's severity and comparing one year with another can prove challenging. One can do so using a variety of measures including the number of fires, acres burned, length of season, structures lost, incident management teams mobilized, or fire-related fatalities. For those in the firefighting community, the number of fireline and fire-related fatalities may be the most critical measure of a fire season's outcome.

So far this year, almost nearing the wind-down stage of the Western fire year, there have been 14 firefighter or fire-related fatalities. This is a higher number than the 10 fatalities that occurred in 2017, the 12 fatalities that occurred in 2016, or the 13 fatalities that occurred in 2015. In 2013, when the Yarnell incident occurred in Arizona, firefighter fatalities reached an all-time high with 34 firefighters losing their lives that year.

The number of acres burned provides a metric both useful to compare years and to quantify severity of a given year. It can also be a long-term indicator of fire conditions and trends on the landscape. As of September 7th, more than seven million acres have burned in 2018. In the past 16 years, we've surpassed the seven million-acre mark eight times, and the nine million-acre mark five times. So far, the peak year for acres burned by wildfire peaked in 2015 at 10.1 million acres. In the 10 years prior to 2002 (1992-2001), we reached 7 million acres on only one year, and had six years with less than 3 million acres burned.

2018 has been an active fire season, with more than 30,000 people assigned to fire incidents during the peak in early August. At this point, 2018 has not yet been one of the "worst." However, with several months left during which California will experience critical fire conditions and fall fire seasons approaching in the east, Lake States, and elsewhere, more acres will burn before the fire year ends. Many California residents likely feel as though this fire season was one of the worst, as more than 1.4 million acres have burned throughout the state, 2,356 homes have been lost to wildfires, and 14 lives have been lost.

According to 2018 statistics for the Northern Rockies Geographic Area alone (which includes Northern Idaho, all of Montana, and the western Dakotas), the Forest Service reported 1,851 fires that burned 99,130 acres, YTD. However, the 2018 fire year is not yet behind us.

We remain cautious and alert as there has not been substantial widespread moisture since July 3rd, fuels are still dry, and we are having new fires every day.

Our Nations Forests and Wildfire

Fire is a natural phenomenon for nearly every forest ecosystem in this country. Fire has shaped the occurrence and distribution of different ecosystems for centuries, simultaneously impacting the human and plant and animal communities in and around those forests. Over the past century, a culture of fire exclusion unfortunately removed the natural role of fire from the public consciousness. When combined with a reduced level of forest management in many areas of the country, fire exclusion led to the build-up of forest fuels to unprecedented levels. Despite our attempts to manage away wildfire, many of our forests are more fire-prone than ever.

What Federal, State and local fire managers and scientists have learned over the decades is that, by managing hazardous fuels, natural resource agencies can mitigate both the severity and impacts of wildfire. Experience has now shown us that wildfire suppression without proactive forest management is unlikely to result in the least amount of wildfire over time because forest fuels continue to build up to the point where wildfires eventually become unmanageable. Consequently, the challenge of land managers is to mitigate the risk to both human communities and ecosystem values both in the short and long-term by implementing a coordinated and science-based approach integrating fuels reduction, fire suppression, and community planning.

Hazardous fuels reduction includes two commonly applied components; prescribed fire and mechanical thinning. Both can have a beneficial impact on smoke emissions from wildfires because they reduce combustible material. We believe that prescribed fire is an essential hazardous fuels reduction tool, that hazardous fuels reduction helps maintain the "investment" we make in working forests, and that we must manage forests and forest fuels in order to reduce the occurrence of catastrophic wildfire.

Hazardous fuels reduction does not prevent wildfire from occurring but can influence how a fire burns. Experience shows that actively managed timber stands often burn with less intensity than those stands with higher fuel loading and often produce less smoke. Active hazardous fuels reduction can create safer conditions for wildland firefighters to conduct suppression activities and may also offer crews opportunities to keep fires smaller.

Wildfire and Air Quality

There exists ample research evidence documenting the air quality and public health impacts of forest fire smoke. Of primary concern is particulate matter (PM), produced from the combustion of woody material. Specifically, particulate matter smaller than 2.5 microns (PM 2.5) is of concern for individuals exposed to wildfire smoke due to the ability of these small particles to penetrate deep into the lungs and respiratory system. PM 2.5 can cause both short-term health effects such as eye, nose, throat and lung irritation, coughing, and shortness of breath; as well as long-term effects on respiration and the worsening of medical conditions such as asthma and heart disease. Air quality impacts from wildfire smoke often hit the hardest in sensitive populations (such as children, the elderly, and those with pre-existing conditions). In addition to human health, reduced air quality from wildfire smoke can impact tourism, recreation, education, and other aspects of community and economic life.

Relatively recent research suggests that the air quality impacts generated from prescribed fire smoke differ from the smoke produced by unplanned wildfires; and those differences are important to recognize. Prescribed fires are timed to occur under conditions in which fire managers have an increased control over fire location, spread, intensity, and other parameters. In Montana, advanced weather forecasting and state-of-the-art smoke modeling, coupled with cooperative engagement between the natural resource agencies in the state and the Montana Department of Environmental Quality's air quality program staff, allow fire managers to tailor ignition locations and times to meet specific smoke management objectives. While each state has differing laws and regulating burning, fire managers work within these parameters and laws with the intent of producing as little smoke as they can, with the intent being to produce as little smoke through prescribed burning to avoid much greater amounts of wildfire smoke in the future.

The US Environmental Protection Agency (EPA) acknowledged, in its rulemaking over the past two years, the benefits of managing smoke production during prescribed fire activity in order to reduce air quality impacts. In both the updating of the National Ambient Air Quality Standard (NAAQS) for PM 2.5 (81 CFR 164, pg. 58010) and the updating of the Exceptional Events Rule (81 CFR 191, pg. 68216), the EPA clearly documents the role of wildfire as an emissions source and the relevance of prescribed fire use and fuels management to reduce the risk of catastrophic wildfire. It is becoming increasingly evident through both research and experience that without prescribed fire and the relatively small amount of managed smoke that comes with it, we are perpetuating the conditions that generate catastrophic fires and resulting air quality issues, while simultaneously putting people and their communities at risk.

Let us not forget about the impact smoke exposure has on our wildland firefighters. Last Friday, September 7, 2018, the Joint Fire Science Program (JFSP) released a Final Report entitled "Wildland Fire Smoke Health Effects on Wildland Firefighters and the Public." The report examined smoke exposure concentration data for wildland firefighters to estimate the health risks specific to prescribed and wildland fire smoke. Findings suggest that wildland firefighters may be at more risk for smoke exposure and health-related impacts when working on wildland fires compared to prescribed fires.

However, scientists cite the need to do more research to understand the true extent of health-related impacts to firefighters who work on both prescribed and wildland fires. This need speaks to the importance of maintaining JFSP funding. Current JFSP lines of work include fuel treatment effectiveness, post-fire recovery, smoke management and air quality, fire risk, and data and software integration frameworks for decision support. These lines of work are of importance to agencies who use the results of JFSP funded research both to inform our management and to help us inform and influence our citizens.

Montana Perspective

Although the 2018 fire year in Montana has been, thankfully, relatively moderate in comparison to other years, over time, fires have been growing larger and more intense throughout the state; and our average fire season is 40 days longer than it was 30 years ago[4]. According to the Northern Rockies Coordination Center (NRCC), the 2017 fire year was our largest on record since 1910 with over 1.2 million acres burned statewide and two fire-related fatalities. Using data from NRCC, we can determine that from 1998-2007 an average of 22,828 acres burned per year in Montana. From 2008-2017 the average increased 15-fold with up to 349,598 of wildland acres burning per year. Data also shows that Montana went fifteen consecutive years without a 100,000-acre fire year, from 1997-2010, but had two 1,000,000-acre years in the 2010-2018 period.

With larger and more severe fires comes more smoke. According to Montana Department of Environmental Quality (MT DEQ), the air quality standards for particulate matter have been exceeded 579 times over the past 11 years, with 214 of those occurring in 2017. During this year, particulate matter concentration increased 17-fold from the air quality standards maximum level.

Similar to many forests within our region, forest conditions in Montana are at increased risk of wildfire. Large portions of the state's forests have succumbed to insect epidemics, historically fire-tolerant forests have been replaced by tree species less adapted to fire, and fuel loadings have increased across large portions of our forests, increasing the severity and intensity of wildfires. Due to changed conditions, a disproportionate amount of Montana's fire-adapted forests is at significant risk of wildfire. Today, over 85 percent of Montana's forests are outside of their historic vegetative condition and are therefore at elevated wildfire hazard potential.

Forest conditions and events resulting from those conditions affect all Montanans regardless of whose land they take place on. Insects outbreaks,

---

[4] Freeborn, P. H., W. M. Jolly, and M. A.Cochrane (2016), Impacts of changing fire weather conditions on reconstructed trends in U.S. wildland fire activity from 1979 to 2014, J. Geophys.Res. Biogeosci., 121, doi:10.1002/2016JG003617.

wildfires, smoke, drought, and a changing climate do not recognize ownership boundaries. According to the Montana Wood Products Association, out of 23 million acres of forested land, Montana has 19.8 million acres of productive, non-reserved timberland. 61 percent of this land is National Forest; 25 percent is non-industrial private forest and tribal; 5 percent is industrial private forest, industry; 5 percent is state; and 4 percent is managed by the Bureau of Land Management. Although these various ownerships operate under unique missions, goals, and objectives, by working cooperatively to co-manage fire risk we can mitigate fire severity on Montana's forested landscape and reduce the smoke-related impacts to people and communities. This is especially important on federal lands which account for large areas of western states and on which fire management and fuels treatment have direct implications for adjacent state and private lands and/or communities.

In Montana, we have a long history of working together to resolve intractable land management issues facing our state. Montana has been engaging in regional and statewide efforts to renew our collective commitment on working together with other key stakeholders to reduce the risk of wildland fire and smoke-related impacts.

A. Western Governors' Association

Last summer, the Western Governors' Association (WGA), chaired by Montana Governor Steve Bullock, released the *National Forest and Rangeland Management Initiative Special Report*. This report represents a multi-state, bipartisan collaborative perspective on promoting health and resilience of forests and rangelands in the West and highlights mechanisms to bring states, federal land managers, private landowners, and stakeholders together to discuss issues and opportunities in forest and rangeland restoration and management emphasizing investments in all lands/cross-boundary management opportunities. Recognizing the impacts related to wildfire, the WGA offers specific recommendations to increase hazardous fuels reduction and support advancements in the use of prescribed fire.

Through the bipartisan spirit of the West, and by Governors working closely with Congressional members, recommendations contained in this report address critical issues to increasing restoration on acres at risk from wildfire. The Wildfire Suppression Funding and Forest Management Activities Act within the FY 2018 omnibus bill advanced many of these recommendations. The long-term "fire funding fix"; the GNA amendment to allow for road reconstruction, repair and restoration; and the establishment of categorical exclusions for wildfire resilience projects will facilitate more fuels reduction projects on and off federal lands and allow them to be started and completed more quickly.

B. *Forests in Focus*

In July of 2014, Governor Bullock announced the Forests in Focus Initiative: A Forest Strategy for Montana, to increase the pace and scale of restoration on Montana's forests in order to reduce the risk of wildfire, address forest health issues, and to grow and sustain forest products industry within the state. Included in the Initiative was the dedication of a fund to accommodate cost-shared forest restoration and fuel reduction projects on tribal, state, and private forested lands. Additionally, nearly five million acres of National Forest System land were nominated as "priority landscapes" because those lands were at risk of forest health threats and/or catastrophic wildfires. This designation qualified eligible lands for important Farm Bill tools to accelerate forest restoration. Since its inception, Forests in Focus investments have supported the treatment of over 300,000 acres, production of nearly 200 million board feet of timber, and retention of jobs in the forest products industry sector.

Like many other states throughout the region, Montana is investing in the use of GNA to increase the pace and scale of forest restoration across ownerships and in partnership with key forest stakeholders. This, along with increased investments in stewardship on private non-industrial forest lands, supporting policies and tools that allow agencies to get more work done on the ground, and developing capacity for local governments to accomplish work in and around their communities, we endeavor to decrease wildland fire risk on a much broader scale than we are currently, thereby having a net-positive effect on wildfire smoke-related impacts to Montanans.

C. Use of Prescribed Fire in Montana

In Montana, land managers use prescribed burning as a tool to mitigate the severity of wildland fires by reducing the build-up of flammable fuel in our forests. Prescribed burning also helps maintain biodiversity, and regenerates vegetation. According to the NRCC, thus far in 2018, the State of Montana, as well as its federal and local partners have implemented prescribed burns on 28,049 acres compared to the ten-year average of 37,352 acres. According to the *2015 National Prescribed Fire Use Survey Report* released by the Coalition of Prescribed Fire Councils, out of 17 western states, Montana burns less than average compared to nine other states within the region. Land managers within the state know we need to be doing more with prescribed fire. As the report's nationwide findings state, we find that our main challenges are weather, conditions of fuels during open burn windows, and capacity rather than air quality issues. However, we also acknowledge that, as we look for ways to increase the use of prescribed fire within the state, the public's acceptance of prescribed fire smoke will likely become an issue.

However, air quality issues have not yet presented a major impediment to the use of prescribed fire in Montana according to many land managers. This may be in part to the development of a progressive model to regulate prescribed burning in a collaborative and cooperative manner. The Montana/Idaho Airshed Group (Group) is dedicated to the preservation of air quality in Montana and Idaho and was initially formed to reduce the impacts of smoke from prescribed burning on Montana and Idaho communities. The Group is run by and for major burners to coordinate burning activities and streamline engagement with MT DEQ and County public health officials. The Group's intent is to minimize or prevent smoke impacts while using fire to accomplish land management objectives. The Group is composed of state, federal, tribal and private member organizations. The Group jointly uses an Airshed Management System database to coordinate burning through the Smoke Management Unit. The Group tracks all planned burns and communicates this on behalf of the burners to regulators through one centralized position at the Smoke Management Unit. This allows for greater coordination among burners and air quality regulators by having one person communicating with the regulating agencies. If burning in Montana, members must have an annual air quality major outdoor burning permit issued by the MT DEQ. Additionally, members are required to comply with local air pollution control agency and/or a fire safety outdoor burning permit. In short, the Group assures coordinated compliance with regulatory agencies, and results in more prescribed burning taking place on the ground that complies with air quality standards. According to MT DEQ, over the past 11 years, there have been only 4 instances when prescribed fire exceeded air quality standards for particulate matter, compared to the 579 instances for wildfire. This success is largely due to the coordinating burning efforts of the Airshed Group. Lastly, in the Summer 2018 *Ecosystem Workforce Program Working Paper* "Prescribed Fire Policy Barriers and Opportunities," authors recommended the Montana/Idaho Airshed Group as a model for other states to follow.

Conclusion

Thank you for the opportunity to appear before the Committee today on behalf of the National Association of State Foresters and the State of Montana. As State Foresters, we believe that, to avert a forest health and wildland fire crisis, we need to be doing significantly more hazardous fuels reduction throughout the country. We are working towards this goal and prescribed fire represents a critical tool necessary to accomplish it. Hazardous fuels reduction and prescribed fire treatments decrease fuel loading in the forests so that when wildfires inevitably occur, they burn with less intensity and reduced spread, burn for shorter periods of time, and produce fewer smoke impacts on firefighters, citizens, and communities.

***

Mr. Shimkus. The chair thanks you.

The chair now turns to Mr. Collin O'Mara, president and CEO of the National Wildlife Federation.

You are recognized for 5 minutes. Welcome back.

STATEMENT OF COLLIN O'MARA

Mr. O'Mara. Thank you, Mr. Chairman, Mr. Tonko, Chairman Walden. You have my prepared testimony.

I actually want to have a – kind of a real conversation today, because I think – you know, I really appreciate the topic and actually focus on the health consequences. But the external debate on this issue has become a little ridiculous, right?

Like, one side is saying it is all about logging and, frankly, just kind of cutting everything down. The other side is saying it is all about climate. We need to actively manage and we actually have to address the climate stressors that are causing this system. And this is a more complicated conversation. It doesn't fit into the normal kind of right-left debate.

You are going to have almost unanimous agreement on this panel on 80 percent of the recommendations. I mean, that doesn't happen before this committee all that often, having been in the room with the chairman a handful of times.

One of the missed opportunities in the fire funding fix this year, and I am so grateful because so many of you played a constructive role in it, was delaying the funding for 2 years, to not have it take effect till 2 years. I mean, we need that money in Oregon right now. We need that money in California right now.

And I get it. But leadership, when they jumped in, they didn't listen to some of you, and Congressman Simpson and others. It is billions of dollars of potential – of missed opportunity to do restoration work.

And, look, I mean, the appropriation minibus is already moving. It is probably too far down the line. But there has to be a way to get a slug of money, because the Forest Service is basically out right now. I mean, they have hit the caps they would have hit end of the month. And if we don't get these projects on the ground, we are going to continue to have more and more of this kind of restoration deficit, if you will, that we are trying to undo. Because we have basically starved ourselves for 40 years, right? I mean, at least the last 25.

And you are talking about a lot of funding. There is great reports. There is a great one just put out by Oregon State looking at how to get more prescribed burns on the ground.

Look, there are things we need to do, like making sure that the ambient air quality standards aren't overly prohibitive and making sure that we are accounting for the impacts of prescribed burns in a way that is actually rational, and not discounting natural kind of anthropogenic emissions in a different way than we are treating manmade ones, especially if the manmade ones are going to save us 90 or 100 percent of emissions compared to the alternative.

But most of the problem here is actually funding in collaboration. And Secretary Perdue put out a great report just a few weeks ago talking about shared stewardship, talking about how to use some of these tools that all of you put together in the last fire funding package and actually trying to get more projects on the ground. And there are things around good neighbor provisions that we absolutely have to fund. There is stewardship contracting provisions we have to fund. There is some mechanical issues that we could work through and actually use your help trying to make sure that the good neighbor provisions have the right accounting behind them so States are incentivized to do the work.

I mean, there are some things – there are some additional tools that folks like Congressman Westerman is working on, like Chairman Barrasso, and Senator Carper, some additional little tools on the management side.

But this is like one of those conversations, like, let's not score points on it, right? Like, folks are hurting right now. I talked to my friend who is an air director up in Oregon right now. They are trying to rewrite their smoke plans right now. It is a good collaborative process. It is a few years too late. I would have liked to have seen it a few years ago.

I mean, the leaders that you have on this panel actually have a lot of the solutions. And so I am like if you do the talking points, you can't have this conversation without talking about climate. You got dryer soils. You got less snow pack. You have warmer temperatures. I mean, the fact that it is not a – I am going to steal your line, I apologize, that it is no longer a fire season, it is a fire year.

This is a serious conversation. And, you know, I know there is a lot of other votes going on, but not a lot of folks are here right now. And so I would just – I would encourage – if there is folks that want to have this conversation in a real way – and it is not just the E&C. I mean, it affects Natural Resources. We obviously have jurisdictional issues all over the place.

We got to fix the funding issue. That is the first thing. We have to figure out – we have to figure out some of these collaborative measures and how we basically bolster the collaboratives in a big way, because the collaboratives are the way to get good products on the ground. There is huge opportunity there. And there is some commonsense things that could be fit into the farm bill.

Advancing prescribed burns in a smart way, and there is some guidance – we don't actually need to change the Clean Air Act, but there is some guidance coming out of EPA related to how they actually measure different types of emissions that have to be fixed. I think Administrator Wheeler could get this done. I think, frankly, Gina McCarthy would have agreed with him on some of these things. This is one of those areas, again, that it is not particularly partisan, and frankly, getting those products on the ground.

Because right now, it is easier to try to respond after the fact than it is to actually do the prescribed burn on the front end. Because it is just a headache. The level of review that is necessary to do it is complicated. These folks do it better than most places. The folks in the southeast are probably doing it the best right now.

But there is models there that we have to figure out how to actually get off the ground, because the scale of restoration that we need is massive. I mean, we are gone from doing, you know, a few million acres here. We need tens of millions of acres of your active management across the board. This is a big conversation we need to have.

We can't have this conversation without talking about acting on climate. I know it is a partisan issue. It shouldn't be. We need to figure out ways to reduce emissions, because they are heating up these systems and making them worse.

There is a big oversight role for all of you too. I do worry about the fire funding fix when it kicks in. The extra money needs to go towards restoration. It needs to go towards active restoration, active management. It can't just go to other programs. That is going to require some oversight, because the way the language is written, it doesn't quite do that.

And then finally, I would encourage, especially folks in the west, to try to figure out ways to get more members out to see the impacts. Because right now – I mean, I spend a lot of time in the west. I don't think folks can fully appreciate the level of devastation in the southern California airshed, in these States. I mean, breathing the soot for day after day, this is a big issue. And at a time when we are preparing for massive hurricanes, this is the time for serious people. And I would love to work with all of you, because as, you know, the great American poet Elvis Presley said, you know, a little less conversation, a little more action.

Thank you.

[The prepared statement of Mr. O'Mara follows:]

TESTIMONY OF COLLIN O'MARA, PRESIDENT AND CEO OF THE NATIONAL WILDLIFE FEDERATION, BEFORE THE UNITED STATES HOUSE OF REPRESENTATIVES, COMMITTEE ON ENERGY AND COMMERCE, SUBCOMMITTEE ON ENVIRONMENT, HEARING ON AIR QUALITY IMPACTS OF WILDFIRES: MITIGATION AND MANAGEMENT STRATEGIES, SEPTEMBER 13, 2018

Thank you to Chairman Walden, Ranking Member Pallone, Subcommittee Chairman Shimkus, Ranking Member Tonko, and the members of this subcommittee for convening this important hearing on a critical issue facing our communities, public health, and wildlife.

My name is Collin O'Mara and I serve as the president and CEO of the National Wildlife Federation—America's largest conservation organization, comprised of 51 state and territorial affiliates and more than 6 million members who are committed to ensuring that wildlife thrive in our rapidly changing world. Prior to joining the Federation, I served as Secretary of Natural Resources and Environment for the State of Delaware, where I oversaw forest management efforts on Delaware's State Parks and State Wildlife Refuges, as well as coordinated with our State Forester on projects to improve the health of our State Forests.

Healthy forests underpin a healthy economy and help protect clean water, clean air, our climate, and essential ecosystems. Yet, the ability of our National Forests to do so has been jeopardized by inadequate restoration and management and escalating climate impacts, both of which exacerbate the threat and consequences of increasingly intense wildfires.

Across our country, wildlife and human communities are living alongside the threat of wildfire. For high-risk communities that have not burned, residents fear evacuation orders and the ominous red glow fire casts on the horizon. For those that have experienced modern megafires, they are rebuilding or helping neighbors recover — even as the threat of future fires looms.

Over the past three years, more than 25 million acres of U.S. forests have burned, including more than 10 million acres in both 2015 and 2017. We are on pace to near 10 million acres burned again this year with more than 8.1 million acres burned this year-to-date. Right now, 84 wildfires are actively burning across 12 states from Alaska to Arizona, currently covering more than 1.4 million acres, posing significant risks to people, property, and wildlife. We honor the brave firefighters and offer our deepest condolences for those who have lost friends and family.

In our report issued last year, *Megafires: The Growing Risk to America's Forests, Communities, and Wildlife*, the National Wildlife Federation demonstrated how these catastrophic fires are the leading edge of an emerging, tragic trend. Wildfires have been burning more intensely and frequently than in previous decades. These increasingly large, fast-spreading, intensely hot fires fueled by decades of fuel loading, fire suppression efforts, and climatic changes like decreased precipitation, warming temperatures, and increased pest problems — have fundamentally altered the natural fire cycle and reduced the resilience of forest landscapes.

Wildfire is a natural, and even healthy, phenomenon for forests and wildlife habitat, but this new normal is anything but for millions of Americans and the forests central to our wildlife heritage.

Last year, we saw smoke so thick in places, like Missoula, Montana, and Los Angeles, California, that residents couldn't see across their front yards. Again today in Southern California, Eastern Oregon, and other states, children and the elderly are choking on smoke and need filters to simply breathe. These situations, particularly tragic for low-income families and those on fixed incomes, are becoming commonplace in our age of escalating megafires.

The unhealthy levels of fine particulate matter (especially PM 2.5 and PM 10) and toxic fumes (such as heightened ozone levels caused by the reaction of oxides of nitrogen, volatile organic compounds, and carbon monoxide) exacerbate risks of asthma, pneumonia, and other respiratory and cardiac problems.

In some cases, the increase in air pollution due to wildfires is undermining years of progress reducing air pollution from power plants, industrial facilities, and vehicles. Studies have shown that the annual carbon emissions released by U.S. wildfires can range from 160 million to 290 million tons, which is the equivalent of are equivalent to the emissions of 34-63 million passenger vehicles or 2-4% of total U.S. carbon emissions.

The National Wildlife Federation firmly believes that collaborative, science-based active restoration of our forests, including ecologically-sound controlled burns and mechanical treatments, is absolutely necessary to improve forest health. We believe providing sufficient funding, enhancing collaborative forest management, advancing ecologically-sound prescribed burns, acting on climate, and ensuring implementation oversight are all necessary components of a comprehensive strategy. Here are our specific recommendations:

Providing Sufficient Funding

This March, Congress took a significant step forward to address the problem of wildfire by passing the fire funding fix. Unlike other disasters, such as hurricanes, wildfires have been ineligible for disaster assistance under the Stafford Act and thus the Forest Service has been forced to raid other accounts, such as forest restoration programs, to fund fire response activities. This ends up reducing the pace and scale of restoration efforts that could have otherwise reduced fire threats in future years. The fire funding fix adopted this spring eliminates this uncertainty by capping the amount of money that the Forest Service has to spend of fire response, before having access to emergency funding. This should free up significant resources for proactive forest restoration efforts in future years.

We at the National Wildlife Federation were proud to champion this effort alongside Members of both parties. Unfortunately, the funding component of the final compromise does not take effect until Fiscal Year 2020, which provides no funding relief for this year's catastrophic fires.

We urge Congress to fully fund and implement this common-sense strategy as soon as possible, while continuing to provide sufficient annual appropriations the federal land management agencies need to manage our forests and respond to wildfires.

Enhancing Collaborative Forest Management

A flexible, collaborative decision-making process is key to minimizing the threat of megafires. The U.S. Forest Service recently released a new plan, "Toward Shared Stewardship Across Landscapes: An Outcome-Based Investment Strategy," for managing forests more effectively to reduce fire risks. Built upon the new funding and management tools Congress provided the agency alongside the fire funding fix, the Shared Stewardship strategy represents an important start to improving collaborative, landscape-scale management of our National Forests. That said, we also see greater opportunities for USFS to integrate climate-smart adaptation practices into all restoration projects that improve forest resilience, improve wildlife habitat, and increase carbon sequestration volumes.

Building upon the fire funding fix and additional tools provided by Congress to restore the resilience of our forests and reduce risks from future fires, the Farm Bill conference committee and the negotiators of the Interior Appropriations bill are currently considering additional proposals to improve collaborative forest management.

If Members want to explore additional tools, it is clear the best path forward is the bipartisan approach that resulted in Congress recently passing the fire funding fix. Promising proposals with potential bipartisan support include:

- Reauthorizing the Collaborative Forest Landscape Restoration Program.
- Adding lodgepole pine restoration projects in Fire Regime IV to the Healthy Forest Restoration Act categorical exclusion for priority projects, and to the list of projects that can receive the expedited, focused analysis of the proposed action, no action, and a possible third alternative.
- Reauthorizing appropriations to treat insect and disease infestations.
- Incentivizing forest management projects under the Good Neighbor Authority by ensuring receipts from any timber sold through these projects can contribute to covering the costs incurred by non-federal project partners, and that any receipts in excess of costs are available for the Forest Service to spend on other Good Neighbor Authority or other forest restoration projects on National Forests in the same state.
- Expanding Good Neighbor Authority to include Tribes and counties with sufficient capacity to conduct work in accordance with adopted forest management plans.
- Passing the Timber Innovation Program, currently in the Senate Farm Bill, to create new markets for wood byproducts from forest restoration projects.

Advancing Ecologically-Sound Prescribed Burns

Though it may seem counter-intuitive, increasing prescribed burns is one of the most effective tools to improve forest health and reduce long-term adverse public health impacts, especially when combined with strategic efforts to reduce fuel loads in ways that do not harm biodiversity. While some debate has emerged about the health tradeoffs of engaging in strategic prescribed burns as a key management strategy to mitigate long-term fire risks, studies have consistently shown that prescribed burns emit 10 times to 100 times less particulate matter than wildfires. USFS and States take significant precautions to reduce the health impacts from prescribed burns, such as establishing hourly and daily PM 2.5 limits, as well as specific plans to support at-risk populations, but it should not be lost that the health impacts of uncontrolled wildfires are on average 90-99% worse than prescribed burns.

We agree that funding, interagency collaboration, and insufficient capacity are the most significant barriers to increasing the utilization of prescribed burns—obstacles that the fire funding fix should help alleviate.

We do not believe that changes are necessary to the Clean Air Act to allow for more prescribed burns, but we do believe that EPA guidance and state-level policies could be more supportive and disincentives could be removed.

For example, there is a perverse incentive whereby emissions from prescribed burns ("anthropogenic ignition") are included in the calculations to determine whether a state is in attainment of the National Ambient Air Quality Standards, but wildfires ("natural ignition") are regularly excluded, despite typically emitting 90-99% more pollution. This is exactly backwards and unwittingly dis-incentivizes states from taking proactive measures to reduce risks of more-polluting megafires. Instead, EPA should account for all emissions and prioritize the granting of wildfire accounting exceptions to those states and communities who have ecologically-sound and landscape-scale fire programs.

Further, the federal government could play a much more proactive role in bringing stakeholders together to advance best practices to accelerate the adoption of prescribed burns. The USFS Shared Stewardship Strategy takes important steps in this direction and we are working with Secretary Perdue to implement its recommendations.

In addition, states should be encouraged and incentivized to continue to improve their Smoke Management Plans and implement best practices that protect public health, such as flexibility for communities that take measures to protect vulnerable populations, while encouraging greater collaboration and resourcing to ensure that prescribed burns occur at appropriate times and scales.

Acting on Climate

No conversation about improving forest health is complete without confronting the changing climatic conditions that are exacerbating megafires, especially less precipitation, drier soils and vegetation, and warmer temperatures. We urge Members to collaborate across party lines to advance bipartisan solutions that reduce greenhouse gas emissions through the accelerated adoption of land management practices that increase the carbon sequestration capacity of natural systems, like forests, grasslands and wetlands, as well as from fossil fuel sources through market-based mechanisms (e.g., price on carbon).

Ensuring Implementation Oversight

Congress should prioritize making sure the Forest Service effectively utilizes the additional funding and management tools recently provided to USDA. The Committee should also ensure the agencies is applying the freed-up funding and policies in a climate-smart, wildlife-friendly manner that improves the resilience of forest habitats to disturbances and threats and simultaneously reduces overall adverse impacts to air quality.

Conclusion

We appreciate the Committee focusing on identifying solutions that would reduce these health consequences of escalating megafires. We encourage the Committee to pursue the numerous bipartisan, science-based solutions that are within reach. We believe that a comprehensive approach, including providing sufficient funding, enhancing collaborative forest management, advancing ecologically-sound prescribed burns, acting on climate, and ensuring implementation oversight, will benefit people and wildlife alike. Thank you for the opportunity to testify on this critical issue.

***

Mr. Shimkus. The gentleman's time expired.
Defending my colleagues here, I think this is actually a pretty good turnout. We do have a bill on the floor. We do have a Health Subcommittee hearing upstairs. So this is not bad, so –

The Chairman. Mr. Chairman, you might point out that this is a subcommittee too. When we are in full committee, all these seats are filled, as they were this morning. So just for the audience.

Mr. Shimkus. I would agree.
Reclaiming my time. The chair now recognizes Mr. Boggus, the State forester and director of Texas A&M Forest Service, on behalf of the National Association of State Foresters.

STATEMENT OF TOM BOGGUS

Mr. Boggus. Thank you –
Mr. Shimkus. You are recognized for 5 minutes.
Mr. Boggus. – Chairman Shimkus and Ranking Member Tonko and Committee Chair Walden. I am glad all of you are here. And I am glad to be here, so – to talk about this important issue today on air quality and wildfire.

My name is Tom Boggus, and I am the State forester and director for the Texas A&M Forest Service. I am here to testify on behalf of the National Association of State Foresters, where I serve on the – member of the Wildland Fire Committee, as well as the past president of the Southern Group of State Foresters, which represents the 13 southeastern States.

I have spent 38 years in forestry and fire, and I am honored to share some of that experience with the subcommittee today.

The NASF and the regional associations like Southern Group represent the directors of the Nation State forestry agencies. We deliver technical and financial assistance, along with fire and resource protection, to more than two-thirds of our Nation's 766 million acres of forestland. We do this with critical partnerships and with investments from the Federal Government, including U.S. Forest Service State and Volunteer Fire Assistance Grants, which provide equipment and training to the firefighters who respond to State and private land where over 80 percent of our Nation's wildfires begin.

This has been a heck of a year across the country. You have heard that. And Texas was no exception. We had over 8,000 wildfires burning over half a million acres so far in 2018. The fire activity impacts responders at local, State, and national levels.

The first impact is to communities. And what many people don't understand and realize is that, in Texas, 75 percent of our wildfires occur within 1 mile of a community. Most of these fires, historically 91 percent, are suppressed by the local responders. The other 9 percent, when their capacity is exceeded, require local, State, and often national resources to control.

Wildfires affect us all. I don't care whether you are rural or urban, local or State or national. At the State and national level, demand to respond does not go away. And you just heard from my colleague here that in the wildfire community, we have quit using fire season and we started using fire year, because it is much more accurate. Because there is a wildfire season somewhere, and wildfires are happening somewhere across America at any time.

Fire has always been a natural part of the ecosystem in Texas, in the south, and, really, a lot of parts of the country. However, for many reasons, wildfires have become increasingly detrimental to the forests and communities around them, including the generation of catastrophic amounts of air pollutants. That is why we are here today.

So what can we do to address the massive amounts of wildfire smoke? My State forester colleagues and I put a great deal of emphasis on proactive prescribed burning. During the times of year when fire risk is low, you have already it, where fire size and smoke emissions and community notification can be managed effectively as compared to an unplanned or an often catastrophic wildfire.

In the southern part of the country, we have a long history of getting prescribed fire accomplished on the ground. We have formed a fire management committee in the States consisting of a fire director from each of the 13 States, and we work together on shared practices, best management practices. For example, we created the Southern Wildfire Risk Assessment Portal, or SouthWRAP. And it is especially important in an urbanizing State like Texas.

We build and maintain strong partnerships with landowners and local governments in implementing partnerships with State environmental quality agencies, Federal land management agencies to get prescribed burning done and forest management done collaboratively.

In Texas, unlike the west, 94 percent of our land is privately owned, and prescribed fire is primarily conducted by private landowners. Texas is a big and diverse State. And the reasons for conducting prescribed burning are just as diverse as our geography.

We recently developed a State smoke management plan to provide best management practices for our landowners and these cooperators and certified burners. The plan provides resources for these professionals to utilize in order to minimize the smoke from their prescribed burns.

Environmental regulations such as air quality are under the authority of the Texas Commission on Environmental Quality, the TCEQ. Now, they are also – that is a great conversation in the education process, but they are a great partner, and they understand and have said last week at a hearing in the State that we need more fire on the ground in Texas and more prescribed fire and not less.

So once again, I want to thank you for this opportunity to testify and appear before you. I look forward to answering any questions. And if I can share more expertise that we have in Texas and the south related to wildfire, hazardous fuel reduction, and prescribed burning.

Thank you, sir.

[The prepared statement of Mr. Boggus follows:]

Testimony of Tom Boggus, Texas State Forester, on Behalf of the National Association of State Foresters, Submitted to the U.S. House Committee on Energy and Commerce, Subcommittee on the Environment, for Hearing on Air Quality Impacts of Wildfires: Mitigation and Management Strategies, September 13, 2018

Good morning, Chairman Shimkus, Ranking Member Tonko, and Members of the subcommittee. My name is Tom Boggus, State Forester and Director of the Texas A&M Forest Service, as well as an active member of the National Association of State Foresters' (NASF) Wildland Fire Committee and past President of the Southern Group of State Foresters (SGSF). I appreciate the opportunity to speak with you today and submit written testimony as the Committee considers the significant impacts of wildfire smoke on citizens and communities across the country, as well as the preventive role prescribed fire and hazardous fuels reduction can have in mitigating smoke impacts.

The NASF represents the directors of the state forestry agencies in all 50 states, eight territories, and the District of Columbia. While an independent organization, the SGSF represents 15 of those same State Foresters from across the South, Puerto Rico and the US Virgin Islands. State Foresters deliver technical and financial assistance, along with protection from wildfire and protection of forest health, water, and other ecosystem services for more than two-thirds of our nation's 766 million acres of forests. Through the State Fire Assistance (SFA) and Volunteer Fire Assistance (VFA) programs, state agencies equip prescribed fire managers and wildfire initial attack resources for state and private lands where over 80% of the nation's wildfires start. In addition, State Foresters have a critical role in maintaining healthy forests and minimizing wildfire risk through a variety of management techniques, including preemptive hazardous fuels reduction through tree thinning and prescribed fire.

While the duties of state agencies vary from state to state, all share common forest management and protection missions and most have statutory responsibilities to provide wildland fire protection on all lands, public and private. As such, we are intimately aware of the increasing occurrence of wildland fire and associated smoke impacts in nearly every state. State forestry agencies also work closely with our federal partners in managing complex multi-jurisdiction landscapes, both in wildfire response and in preemptive prescribed burning. As we often say "fire knows no borders," and thus aim to carry out management and planning across ownerships.

Summary of Annual Fire Activity

The current fire season that is still very active has been one of the most newsworthy in recent memory, with numerous incidents making national news. As of September 10, over 47,000 fires have burned more than 7 million acres across our country since the beginning of 2018[5], with significant fire activity still expected before the year is out in California and parts of the southeast. Roughly 75% of those fires have been on state or private land, and thus under the response jurisdiction of State Foresters and local rural/volunteer fire departments.

In Texas, we have seen 8,045 wildfires burn 518,074 acres so far in 2018 – with 9,647 homes reported as saved from wildfires and 108 lost. This fire activity directly impacts responders at the local, state and national level. The first impact is to the communities and local responders across our state – with 75% of Texas wildfires occurring within one mile of a community. While most of these fires (historically 91%) are extinguished by local resources – the other 9% become the large, multiple-day incidents that

---

[5] https://www.nifc.gov/fireInfo/nfn.htm.

account for roughly 2/3 (63%) of the acres burned in Texas – and require local, state and national resources to control.

Wildfires affect us all – rural & urban areas as well as at the local, state and national levels.

At the state and national level – the demand to respond does not go away. The wildfire community has moved from using the term "fire season" to "fire year," as wildfire is now a threat somewhere in our nation at all times of the year, and in places the local threat is year-round. Over the past 12 months, there has been virtually no area of the country immune from wildfire incidents and the associated smoke impacts. This year has been particularly noteworthy, in that fires and their impacts have not been localized to the forest-based communities most experienced with living with fire. Large cities, often far from the forests on fire, have experienced significantly reduced air quality, impacting human health, community events, tourism, recreation, and much, much more. The issue of how to manage smoke from wildfires is an increasingly essential one to address.

Our Nations Forests and Wildfire

Fire is a natural phenomenon for nearly every forest ecosystem in this country. Fire has shaped the occurrence and distribution of different ecosystems for centuries, simultaneously impacting the human and natural communities that live in and around those forests. Over the past century, a culture of fire suppression has unfortunately removed the natural role of fire from the public consciousness to varying degrees in different regions; however, when combined with a reduced level of forest management in many areas of the country this culture has also led to the build-up of hazardous fuels to historic levels. Our forests are currently more fire-prone than ever.

What Federal, State and local fire managers as well as scientists and researchers have learned over the past decades is the critical role of hazardous fuels management in mitigating wildfire impacts. Solely focusing on wildfire suppression and ignoring proactive forest management does not lead to the least amount of fire in the long run; the fuel continues to build up to the point where eventually wildfires become unmanageable under initial attack. The task for wildfire managers is to manage the risk to communities and ecosystem values in both the short-term and long-term by implementing a coordinated and science-based program of fuels reduction, fire suppression, and community planning. Our forests will inevitably burn, the task is to figure out how this phenomenon can occur with the least impact, including from wildfire smoke, on communities.

Wildfire and Air Quality

The air quality impacts from forest fire smoke have long been scientifically documented. Of primary concern is particulate matter (PM), which is produced from the combustion of woody material. Specifically, particulate matter smaller than 2.5 microns (PM 2.5) is of concern for individuals exposed to wildfire smoke due to its ability to penetrate deep into the lungs and respiratory system. PM 2.5 can cause both short-term health effects such as eye, nose, throat and lung irritation, coughing and shortness of breath, as well as long-term effects on respiration and the worsening of medical conditions such as asthma and heart disease. Air quality impacts from wildfire often hit the hardest in sensitive populations (i.e., children, elderly and those with pre-existing conditions). In addition to human health, reduced air quality from wildfire smoke can impact tourism, recreation, education, and a variety of other aspects of community life.

The differing air quality impacts from prescribed fire compared to unplanned wildfire are important to recognize. One of the keys to prescribed fire for hazardous fuels management is that it is done in seasons and under conditions where fire managers have the ability to control fire location, spread, intensity, and many other parameters. Weather forecasting and state-of-the-art smoke modeling software allow for fire managers to tailor ignition locations and times to meet smoke management objectives. While each state has different laws and regulations around burning permits and number of allowable burn days, fire managers work within these parameters and laws to manage a minimal amount of smoke now in avoidance of the potential for a much greater amount in the future.

The beneficial impact of managed prescribed fire on air quality emissions has been recognized by the US Environmental Protection Agency (EPA) in its rulemaking over the past years. In both the updating of the National Ambient Air Quality Standard (NAAQS) for PM 2.5 (81 CFR 164, pg. 58010) and the updating of the Exceptional Events Rule (81 CFR 191, pg. 68216), the EPA clearly documents the role of wildfire as an emissions source and the relevance of prescribed fire use and fuels management to reduce the risk of catastrophic wildfire. It is becoming increasingly evident through science and experience that without prescribed fire and the small amount of managed smoke that comes with it, we are perpetuating the conditions that generate catastrophic air quality issues and put communities and individuals at risk.

Managing Prescribed Fire for Air Quality in the South

The Southern part of the country has a long history of managing fire in its forests, and in coexisting with the smoke that necessarily comes along with fire, whether it be planned or unplanned ignitions. In 2017, the top eight states in terms of number of acres of prescribed fire carried out were in the South, including my state of Texas with almost 200,000 acres[6]. Those eight southern states accounted for over 85% of the prescribed fire acres in the country in 2017.

The reasons that southern states are able to successfully implement such a large program of prescribed fire work are numerous and vary from state-to-state, but there a few regional commonalities that I would like to highlight.

First is the expertise of our state fire and forest managers, and their diligence in planning. Within the South, we have formed a Fire Management Committee, consisting of the Fire Directors from each of our 13 States, who

---

[6] Source: National Interagency Fire Center, https://www.predictiveservices.nifc.gov/intelligence/2017_statssumm/fires_acres17.pdf.

work together to share best practices and develop tools that support the whole region. One such tool is the Southern Wildfire Risk Assessment Portal (SouthWRAP)[7]. SouthWRAP is the primary mechanism for the SGSF to make wildfire risk information available and create awareness about wildfire issues for the Southern states. It is an excellent tool for establishing the basis for "where" and the justification for "why" subsequent prescribed burning is necessary.

This expertise of our state staff and partners has also been used to generate educational materials to ensure that proper best management practices are followed by those conducting burns, minimizing smoke impacts and helping retain social license to burn. One such example is the Prescribed Fire Smoke Management Pocket Guide, created by the Coalition of Prescribed Fire Councils and the Southeast Regional Partnership for Planning and Sustainability (SERPPAS).[8] The Pocket Guide includes how-to videos on basic smoke management guidelines, fact sheets on specific smoke-related issues, and a comprehensive list of additional resources to help any prescribed burn manager complete their job effectively and efficiently from a smoke perspective.

In Texas, state and federal agencies implement prescribed fire programs, but most prescribed fire is conducted by private landowners. Therefore, my agency is using concepts of the pocket guide to develop materials aimed at private landowners. Recently, we have worked to develop a Texas State Smoke Management Plan to provide best management practices to landowners and cooperators. The plan provides resources for prescribed burn managers to utilize in order to best manage the smoke from their prescribed burns.

A second reason that the South has such a strong record of prescribed burning is that our state forestry agencies have built strong partnerships in many arenas which help facilitate prescribed burning work getting done on the ground. In many of our states we have employees in every county that work with local government and landowners to make them aware of any planned burning and keep community support for such necessary activities.

---

[7] https://www.southernwildfirerisk.com/.
[8] http://smokeapp.serppas.org/.

Our states also work with many partners, NGOs, and federal agencies to accomplish the planning, treatment prioritization, and eventual work on the ground. In the spirit of the Cohesive Wildland Fire Management Strategy, our States and our partners across the region team up regularly to create and restore resilient landscapes with prescribed fire. The Cohesive Fire Strategy provides a framework to minimize the risks associated with wildland fire in the Southeast while increasing the region's fire-resiliency.

Additionally, good communication and partnership with the State environmental regulatory agencies is essential to help them understand the need for prescribed fire, allowing some emissions under good weather conditions in order to avoid massive emissions down the road. One regional example of such efforts is regular "smoke summits" that have been organized with University extension specialists, staff from EPA Region 4, and air quality officials and fire chiefs from the eight southern states in that EPA region[9] to discuss issues around smoke management and air quality. These are the only regional meetings of this type in the country.

Finally, the land ownership matrix in the South, with our forests being primarily privately owned, allows for fewer process hurdles to getting work done. Each of our states, including Texas, has a process in place to permit and track prescribed burns throughout the year. This is important because over 94 percent of Texas is privately owned and the majority of prescribed burning is done by private landowners. Texas is a diverse state from the forestlands in east Texas to the rangelands in central and west Texas. The reasons for conducting prescribed burning are just as diverse as the geography of Texas. Some fire managers utilize it to reduce fuel loading to protect communities, some use it to improve rangeland, watershed quality, and wildlife habitat, and still others use it to manage forest or ecosystem health. This requires a collaborative approach to prescribed fire. Because of this, the statutory authority for prescribed burning was placed under the Texas Department of Agriculture. The Texas Department of Agriculture's Prescribed Burning Board (PBB) regulates the Certified Prescribed Burn Manager Program. The Board is comprised of representatives of state

---

[9] EPA Region 4 includes Alabama, Florida, Georgia, Kentucky, Mississippi, North Carolina, South Carolina, and Tennessee.

agencies, private landowners, and researchers. In Texas there are 64 certified burn managers and there are 11 prescribed fire councils.

Environmental regulations such as air quality are under the authority of the Texas Commission on Environmental Quality.

In other parts of the country that have a majority of federal forest land, the NEPA planning and associated processes for burning projects can be cumbersome, or the staff and resources to carry out the burning can be lacking. Finding new and unique ways to accelerate the pace and scale of prescribed burning and active timber management on federal lands across the country is key to reducing wildfire risk and associated extreme smoke emissions, which is why making use of opportunities like Good Neighbor Authority (GNA) on federal lands is so important no matter what region of the country you are in. One of the essential ingredients for the substantial amount of prescribed burning on the South's primarily privately-owned forest landscape is a commitment by States and partners to hire and train the requisite staff to get the work done. Our federal forests need to be similarly equipped for the pace and scale of active management that faces them.

Conclusion

Thank you again for the opportunity to appear before the Committee today on behalf of the National Association of State Foresters. Managing fire in Texas's forested landscapes is one of the most challenging facets of my job, as it for most of my colleagues in their respective states. As State Foresters, we believe we need to be doing significantly more hazardous fuels reduction all across this country and are working towards this goal as individual members and as an association. I am proud of the work our state fire managers and partners are accomplishing on the ground. Such treatments allow us to put fire on the landscape at times and under conditions that minimize impacts, including smoke emissions. These treatments reduce fuel loading in the forests so that when wildfires inevitably occur, they burn with less intensity, reduced spread and fewer smoke impacts on communities and firefighters.

My colleagues and I look forward to continuing our strong working relationships with the federal agencies, state and federal environmental quality agencies, and other partners, as well as working with Congress to facilitate more good work getting done on the ground.

***

Mr. Shimkus. Thank you very much.

Seeing no other members of the panel, I would like to recognize myself 5 minutes to start the round of questions.

And, Mr. O'Mara, I want to go with you just because of your opening statement. And I think you alluded to a missed opportunity in the minibus. I mean, I know you can fill the space, just briefly tell me, what was the missed opportunity?

Mr. O'Mara. Yeah. In the fire funding fix that was passed in March as part of a big budget deal, there was a provision that was snuck in at 3 a.m. that basically moved it from being 2018 fiscal year to 2020 fiscal year. There is no increased funding through the fire fix for the next fiscal year. So you are not going to have the additional money that you all passed for 14 more months. There is some supplemental money that folks here and, you know, Udall and Murkowski put in, but the actual tool isn't available when we are having these horrible conditions.

Mr. Shimkus. Great. Thank you very much.

Let me go to Ms. Anderson and Ms. Germann. Compare and contrast for me the risk challenges and the environmental quality aspects of a forest fire and the resulting smoke and stuff versus auto emissions in coal-fired power plants.

Ms. Anderson. So in Idaho, we don't have any coal-fired power plants. The next biggest emitter are – you know, we do have quite a bit of open burning. We have agricultural burning, backyard burning, a lot of auto emissions. We don't have a lot of industry in Idaho. So by far, the wildfire emissions are the biggest air pollution source that we just can't manage. We have to react to.

Mr. Shimkus. Ms. Germann.

Ms. Germann. Yes. Thank you. I lack the specifics on any type of coal emissions. But I can say anecdotally, certainly, wildfire smoke is, by far, the largest polluter within the State.

Mr. Shimkus. Great.

Let me go to Mr. Boggus. In your testimony, you say that, and I quote, "Our forests are currently more fire prone than ever." I think Mr. O'Mara may have alluded to that. Some of the opening statements would.

Why do you believe that is the case?

Mr. Boggus. We need more active management. And I think several people on the committee have alluded to that. And I think when you have a built up of fuel – and what we haven't even really talked about is the land use changes that have happened. We have got more people living in and around our forests, but the fuel loads are increasing every year.

Mr. Shimkus. And when you use that terminology for, the fuel load is increasing, what are you referring to?

Mr. Boggus. There is more to burn – available to burn in the woods than there ever has been.

Mr. Shimkus. So, you know, Mr. O'Mara, I am not picking on him. I mean, he mentioned the threat of clear cutting. We are not talking about clear cutting large swaths of ground. We are talking about what?

Mr. Boggus. No. We are talking about active management, forest management of the resource.

Mr. Shimkus. Removing some of those fuels.

Mr. Boggus. Yeah. That doesn't mean harvesting, that doesn't mean thinning. But that means keeping forests healthy. And I have great examples, and we won't have time to get into them, but examples in Texas where a managed forest, even if you have severe drought or you have wildfires, the managed forests bear better and you don't have the damage – catastrophic damage that you do to wildland – wildlife habitat and the resources that you do with unmanaged forests.

Mr. Shimkus. Let me go to Mr. Baertschiger – Senator, I am sorry – for that same question.

Mr. Baertschiger. Well, you know, from a fire science perspective, you always have to remember, you have to have drying of the fuels to a point where they will – can be ignited and sustain ignition, and you have to have ignition. You can have the driest and even huge fuel loadings, and if you don't have ignition, you have no fire.

And so when I talk about the human element, that is something that I have been tracking now for about 10 years of really looking at it. So we are having more and more of these fires that are caused by the human element. And when we get more and more fires, then we spread our resources and we can't concentrate on putting one out because it is kind of like whack-a-mole, you know.

Mr. Shimkus. Of the fires that – I mean, of the fires we are experiencing, percentagewise, how much are natural caused and how many are caused by human intervention, a fire not left, or someone – we have had some intentional fires set.

Mr. Baertschiger. Yeah. When I referred to human cause, I am not talking about an arsonist. I am talking about it can be a power line failure, it can be a chain dragging from a vehicle down the road. It is something that has to do with a human, that we wouldn't have that fire if we didn't have that human element into it. And it is getting close to 9 out of 10 fires.

Mr. Shimkus. Okay. Great.

My time is close to be expiring. I will turn to the ranking member, Mr. Tonko, for 5 minutes.

Mr. Tonko. Thank you, Mr. Chair. And again, welcome, to all of our witnesses. Thank you for your expert testimony.

And, Mr. O'Mara, you made the observation that a great many of us agree about the severity of the problem and the need to move forward. And I am hoping that somehow we can be inspired to come up with solutions that incorporate the professionals that manage these resources in such an outstanding manner.

I want to start with the big-picture question before we get into the specifics on forest management. How important is addressing climate change which we know contributes to conditions that exacerbate the number and severity of these fires for a long-term fire mitigation and our forest management strategy? Mr. O'Mara. I mean, look, I mean, we have to address the underlying stressors of the system long term. You know, those aren't improvements that will happen overnight. There is a lot of things we have to do in the near term. But if we want to have long-term kind of sustainable health, we have to kind of bend the curve on the warming planet.

Mr. Tonko. And in March, Congress passed the fiscal year 2018 omnibus appropriations bill, which included changes to how we fund the United States Forest Service's fire response beginning in 2020. And you alluded to that funding.

Does everyone agree it is important to provide greater funding for more proactive forest management to reduce the risks of these large wildfires?

Mr. O'Mara. Absolutely. I mean, I think the more that we can do, to my colleague's point, about – thinking about private lands, State lands, federal lands – I mean, these are the same landscapes. You know, the ownership might be different, and I do think providing additional funding for certain tools like prescribed burns could be very effective.

I actually don't think that the fix itself is going to end up being sufficient long term, even just given the scale of the – you know, we are talking, you know, another 10 million acres this year probably by the time we are done. I mean, we are escalating in a pretty concerning way. And I think you are going to need more money, frankly, not less.

Mr. Tonko. So is that the additional work that we need to secure here, or is there something more than just the dollars that are required as we go forward with the fix?

Mr. O'Mara. From my point of view, I think there are additional tools that we can provide. I think there are some very important tools that were provided as part of the funding fix in March.

I mean, a few of the ones that, you know, just kind of come to mind, top of mind, is there is like the collaborative forest landscape program that is a very effective tool that is in the current draft of the farm bill, assuming that gets done. There is some things around funding disease and infestation. There is things around Good Neighbor Authority, like you mentioned, making sure that works for everybody, you know, including Tribes, including other partners, counties in some cases that are bigger.

There is some innovation programs for trying to have markets for some of these products, because one of the worries I have is that if we don't create robust markets and trade comes into this, because, you know, a lot of the timber guys are struggling right now because the markets are closing, in China in particular. And so there is a bigger conversation with the economic consequences, making sure they have a place to put this material into good use.

Mr. Tonko. Thank you.

And based on the recommendations of NWF's Megafires report, do you have other suggestions on how Congress and the administration can help reduce the threat of wildfires?

Mr. O'Mara. I mean, I would encourage this committee to convene some of the stakeholders and the agency heads that are involved. Secretary Perdue and his team have done a really nice job, Acting Director Administrator Christiansen and Jim Hubbard.

I think pulling together some folks at EPA and having conversations about how we encourage more prescribed burns and the way they are protective of public health, having some more clear guidance could be helpful. And then also highlighting the success of particularly the Montana-Idaho collaborative event, because I do think that is a model that could be replicated in other places. There is good collaborative in California as well that could be replicated. But we have to elevate these best practices in other places, because we are going to see the impacts get worse over time.

Mr. Tonko. And a few people have mentioned forest provisions included in the House farm bill, H.R. 2, although there have been criticisms that they go too far in undermining environmental laws, including NEPA and the Endangered Species Act.

Do you have any thoughts on those provision?

Mr. O'Mara. Yeah. I am happy to provide additional detail, kind of point by point on them. I think there is a series of them that are very bipartisan. I think there are a few that probably go a little too far in some of the categorical exclusions. You know, we probably should be using the ones that we have right now. I think the one that was passed before was the most important one from March.

And I think, you know, the more that those conversations are being directed by the science, by the experts, the better. But I do think there is a suite of four or five of them that easily could move through this farm bill. And I would love to work with you offline to tell you exactly which ones those are.

Mr. Tonko. Sure. I appreciate that.

With that, Mr. Chair, I yield back.

Mr. Shimkus. The gentleman yields back his time.

The chair recognizes the chairman of the full committee, Mr. Walden.

The Chairman. Thank you again, Mr. Chairman. Thanks to my colleagues for participating in this hearing and to all our witnesses for being here today.

I want to talk about some of the issues that we have run into, some of the data that we have. According to the Georgia Institute of Technology, when they did a study on this, found that wildfires burning more than 11 million acres spew as much carbon monoxide into the air as all the cars and factories in the continental U.S. during those same months. I am sorry, that was California Forestry Association. You are probably familiar with these data points.

And the Intergovernmental Panel on Climate Change, the IPC's fourth assessment report on mitigation said in 2007, quote: In the long term, a sustainable forest management strategy aimed at maintaining or increasing forest stocks while producing an annual sustained yield of timber, fiber, energy from the forest will generate the largest sustained carbon mitigation benefits.

So, basically, healthy green forests sequester carbon. Dead, dying, old ones and ones that burn and re-burn actually emit carbon.

So one of the provisions in the farm bill is something that we do on those other landscapes you referenced, Mr. O'Mara, and that is, after a fire, you harvest the burn dead trees where it makes sense and you replant a new green forest which will sequester carbon.

Is that one of the provisions you oppose – your organization opposes or supports?

Mr. O'Mara. No, no. We have been supportive.

The Chairman. Of the House farm bill provision?

Mr. O'Mara. And the only thing – we just want to make sure we are planting kind of smartly, right, in terms of what is going to be sustainable in the long term.

The Chairman. Sure.

Mr. O'Mara. Oh, no. Absolutely. We – I mean, yes. Absolutely.

The Chairman. Yeah. I mean, because it will be the types of trees for that area and the environment and all that. I mean, we got to be smart about it.

But what I hear, and, Senator, you may want to speak to this, because you both have been on the forest management side and had a career on the forest firefighting side, so you have seen both. Tell me what happens in these fires the second go-around after the trees on Federal ground have not been removed, the burned dead ones. What happens there when a fire breaks out the second time, which often is the case?

Mr. Baertschiger. Well, on Forest Service lands, they are not going to replant after a fire. So when you have the first fire go through, the mortality rate of the live trees is pretty high. The second time or the third time it goes through, it takes out the rest of the trees. So there is no trees to cone out. Cone out means when a tree is starting to die, they will drop cones and reseed and start all over again. But after the second or third burn, there is no trees to do that. And so it changes the entire ecosystem of that forest. You will not have the same forest that you had. And that is what we are seeing in – and the dirt. Yeah. I mean, catastrophic high-density fire.

The Chairman. This is the dirt which remains, which is called ash.

Mr. Baertschiger. Yeah.

The Chairman. And on the second fire, doesn't it make it even harder to maintain any kind of vegetation, frequently, because it burns the soils, it sterilizes the soil so deeply?

Mr. Baertschiger. Our common terminology is it nukes the soil.

The Chairman. Nukes the soil. How far down will it nuke the soil on a bad fire?

Mr. Baertschiger. Just depends how hot it gets. And in southern Oregon, northern California where we have extremely high fuel loadings, in other words, tons per acre, we have a very hot, hot, hot fire. We can have 400-foot flames from some of those fires.

The Chairman. Four hundred feet high?

Mr. Baertschiger. Four hundred feet high, the flames. So depending on the severity, the hotter the fire, the deeper it is going to go into the soil. It can go pretty deep.

The Chairman. Mr. O'Mara, I fully agree with you on the need to solve the fire borrowing issue. I have been an advocate of doing that from day one. It makes no sense. I am told there are statistics that – you know, it costs four to five times as much to fight a fire as it does to do the kind of work you and I agree needs to be done on the forest.

I had somebody in region 6 Forest Service at one point tell me 70 percent of the Forest Service budget for these projects goes into planning, planning and appeals. And it seems like we have got a broken process, then, if all the money is going into the planning and not going to the ground. Do you agree?

Mr. O'Mara. Yeah. And I think there is two issues there.

One is that – I mean, I think there is some redundancy in the planning process. There is some things they could be streamlining. We are not bolstering the collaborative enough. If we have to go through a collaborate process, there should be a way of –

The Chairman. I was a cosponsor of that legislation to do landscape scale collaboratives that we are using in Oregon today.

And I think you said something too about we got to do bigger expanses on these collaboratives, right? Or on the treatment, because we are millions of acres behind.

One of the others provision in the farm bill would extend the categorical exclusions out to 6,000 acres. We have got millions we need to do. Three thousand is currently on the books, but only on certain forests in certain States have certain governors identified certain lands.

So in southern Oregon where the Senator is from, our Governor didn't designate any of those lands. But the provisions in the farm bill in the House would allow for a 6,000-acre CE where you could go in and begin this catchup work. And so I am hopeful we can get that into law.

Our committee – while we want to believe we have complete jurisdiction over every issue on the books in the Congress, and I think we would be better off if we did, doesn't fully have Forest Service jurisdiction, but this is our hook, because what is happening on Federal lands is dramatically, dangerously affecting the health of our citizens, and that is why linking to the air quality is so critical.

Do you want to respond?

Mr. O'Mara. Just one thing. Your point on the carbon emissions, in 20 to 30 percent of the global solution could come from repairing these kind of natural systems.

The Chairman. Absolutely.

Mr. O'Mara. It could be 10 to 15 percent of this country.

When you are talking about the impacts just to the forests for the last few years, I mean you are talking 36 million cars. Right?

This is one of the most potentially bipartisan ways we restore our forests, we reduce emissions. It is a win for everybody.

The Chairman. And you haven't talked about the habitat, the water quality, et cetera, et cetera. My time is expired. The chairman has been very generous. Thank you, Mr. Chairman.

Mr. Shimkus. The chairman is always generous to the chairman. So the chair now recognizes the gentleman from California, Dr. Ruiz, for 5 minutes.

Mr. Ruiz. Thank you very much. And chairman, I agree with you, this is a definite public health concern. And there is two main points: One is that it is a public health concern, just recently in the fires in my district, I had to give a warning on social media to anybody who can smell smoke or see ash, especially vulnerable populations, the older, the young, and people with lung illnesses, that they should be inside in a closed air conditioned unit.

The second main point that this tells us is that these fine particles, particle matter 2.5 microns and substance from a fire in California can be – can travel clear across the country. So whether you are in a fire-prone State or not, it is an American issue and all of our public health can be in gem pardon.

As we sit here today, there are 17 active wildfires burning across the State of California. The ongoing wildfire season has resulted in over 1.4

million acres burned, and the worst is likely yet to come due to climate change. As we know, that climate change can fuel the severity, frequency and the size of wildfires by increasing the duration of droughts, causing long stretches of low humidity and high temperatures, and initiating early springtime melting, which leads to dryer lands in summer months.

So we need to address and recognize climate change and do everything possible, or else we are not being as effective as we can.

In August, the Cranston fire burned over 13,000 acres in my district outside of the community of Idyllwild. This fire exposed the residents of those mountain communities to numerous risks beyond just the flames themselves. While the fire burned, residents across southern California were subjected to increased air pollution as the smoke traveled across the region; these are kids with asthma; elders with COPD; people with pulmonary fibrosis, et cetera, were having more shortness of breath, visiting emergency departments, requiring more intensive care.

The smoke and pollution from wildfires can affect populations far removed from these fires themselves. The fires in California can cause vast clouds of hazardous smoke that can affect the air quality for residents in Arizona, and Nevada and further east.

So wildfires are regional disasters with national implications. And earlier this year, my Wildfire Prevention Act was signed into law, which extended the Hazard Mitigation Grant Program to any fire that receives a fire management assisting grant. Previously, this funding was only available to declared major disasters and not fire. Hazard Mitigation Grant Program funds may be used to fund projects that will help prevent and mitigate future fires. Some examples can include receding construction of barriers, hazardous fuel reduction or reinstalling ground cover. So Mr. O'Mara, can you speak to examples of mitigation projects that can be taken in the wake of wildfires that would be most helpful to preventing a repeat event?

Mr. O'Mara. This is one of those great examples of an ounce of prevention would be worth a pound of cause. There are things you do on the landscape. You talked a lot about prescribed burns, you talked a lot about active management practices, they are ecologically sound, but there is also some common sense. We were building further and further into the wildland urban interface.

Mr. Ruiz. Right.

Mr. O'Mara. And you get folks that are building up into the hills. There is some common sense that we are putting people in harm's way. And I do think there has to be some kind of accounting for that, and making sure we are not putting additional folks aren't in harm's way. It is unfortunate in some cases. These are beautiful places, but we allow people's desire to live in the middle of the woods.

Mr. Ruiz. What are some examples, specific examples that households can do and that we can do as policymakers?

Mr. O'Mara. Sure, there are things on building codes, making sure more fire resistant products and things like that, and some States have done that, or some local governments. There are things in siting that can be incredibly helpful. Making sure climate science is part of your planning process. There are a wide range of things that have people in less harm's way.

Mr. Ruiz. Ms. Germann, as a State forester, can you give examples of how you would use additional hazard mitigation funds to prevent future wildfire damage?

Ms. Germann. Yes, thank you. I can think of several. And I will speak specifically to working on private lands. Any funding that we get through State and private forestry, we are targeting lands within the wildland urban interface to work with landowners to reduce the fields in and around their home, and educate them on things like the home ignition zone. And we are finding that a lot of fires, they also start – homes also burn because of the expanse around the home, if they are not necessarily going to be planting fire resistant material, or shrubs, we try to work with people to educate them on the best type of landscaping that they can have. So it is going to take a couple of things, fuels reduction outside of that home and ignition zone and also work within and around homes.

Mr. Ruiz. It is amazing to see the photos of the houses that were spared because of what they did around their house to mitigate the propagation of fires, it works, it definitely works. I yield back.

Mr. Shimkus. The gentleman yields back. The chair now recognizes the vice chairman of the subcommittee, Congressman McKinley from West Virginia, 5 minutes.

Mr. McKinley. Thank you, Mr. Chairman.

Unfortunately, the ranking member from New Jersey has left. I wanted to thank him for his opening statement, because it gave us – those that are here – a little snapshot of what life could be like after November, if he becomes the chairman, a diatribe of challenging President Trump for everything on climate change. It just shows that such a distraction is going to take place in this committee in the years ahead when we try to deal with all the matters that come before this committee.

And perhaps, it was just meant to be a distraction from the economic insurgence that has taken place across this country. And I appreciate you, Senator from Oregon, that you didn't blame President Trump for one of those nine of ten fought fires being started. He got blamed for waters rising in the oceans, blaming Hurricane Florence. It is just inexcusable, but that is what we are going to see. So it is a little vignette of what we might be able to see in the future.

My question – further with remarks would be, we had some discussion a couple years ago about the CO_2 emissions out in the atmosphere. And I quoted O'Mara, I quoted from Al Gore's book that the largest producer of the CO_2s into the atmosphere is not from coal, it is coming from the deforestation of tropical rain forests. So the idea of what we are seeing in Oregon, California and elsewhere is we are contributing to this. That is why we need to address those problems and solutions so that we are not allowing this uncontrolled burn in our forests and allow that to take place.

Now, I go to West Virginia and there we have the Mon, which is about 1 million acres. Like I say, Mr. O'Mara, with all due respect, it has been groups like yours and others that have prevented the logging in the Mon forest. It is a million acres, and they have only received about $1 million worth of harvest. Think about that: $1 per acre is all they are getting out of that forest. But yet, you go to the Allegheny Forest in Pennsylvania, and it is getting $12 per acre. So we think about what the situation is we have in the Mon. I want to learn from what testimony has been given here, that we may be sitting on something that is a very aging force in West Virginia in the Mon. And it is a tinderbox, because people are preventing us from logging and perfecting the situation that we have in West Virginia.

So I am looking for some guidance as to how we might be able to approach this, because I am afraid we are going to start experiencing the same problem in West Virginia in the Mon that you all were experiencing out west because of environmental groups do not want to have – I have got here, the West Virginia legislature was trying to do some in the State forest, but the environmental groups prevented that from happening.

What advice can you give us for other areas? We have seen the devastation and we have seen the collection that the chairman has of soot from out west, what do – how do we prevent that from happening in the east as well? What would you suggest, any of you? Don't be shy. There is nothing we can learn?

Ms. Germann. Is the question what would we –

Mr. McKinley. What would you recommend? How should we go about this, because the National Forest, because of the environmental movement, is preventing us from thinning that and addressing the problem? We are only get one-twelfth of the wood products out of the Mon that people are getting at other national forests. It is becoming a nursing home for wood.

Ms. Germann. If I may, I think something it that is happening right now, and I think you see it through the panel and Mr. O'Mara and my colleague, Mr. Boggus, we are talking about a lot of the same things. I think there is an opportunity that is happening right now is we are all interested as land managers, and as people who are interested in getting restoration and protecting water quality and air quality. We are wanting to focus on taking a cross-boundary approach. So we call it "All lands, all hands." I think that is something we talk about across the Nation. But we have this opportunity right now to be doing more, but we have to be making sure that we are not only going to be doing more on private lands, we have to have the funding through our agencies for State and private forestry within our State. Other things like Good Neighbor Authority. So it is an excellent partnership between the Federal Government and the States. Working with collaboratives, working with local governments –

Mr. McKinley. Again, those are great ideals, but it is not happening. So Mr. Boggus, what would you suggest? What do we have to do to try to encourage the Forest Service to eliminate these hazards so that we don't experience this same problem?

Mr. Boggus. Well, you have to keep the dialogue open. We are an early adapter, Texas is an early adapter for the Good Neighbor Authority, where you have these agreements – even before there was Good Neighbor Authority, we went into agreement with our State – our national forest folks in Texas, and to help them with prescribed burning. We had an agreement in 2007 and 2008 for that. Then we had the Good Neighbor Authority, which means the States can help the U.S. Forest Service get management done on their lands. And you all's thank you for the fire fix as has been said before, but that is a great help to us, because a lot of times, the money we have and for reaching and technical assistance and the money that people don't talk about is the State and private funding that comes from you all; the borrowing came from State and private often, and so that is where we can reach out and do more on U.S. Forest Service lands, but also on technical assistance and helping the State and the private landowners across the State, which we heard was most of the land. Most of the forest land in this country is on – what you are talking about in the east and the south is on private lands. And those folks need technical assistance.

So these programs like stewardship, Urban and Community Forestry and the Good Neighbor Authority help us put things not just in a plan, but put them on the ground and manage and make our forest healthier.

Mr. McKinley. Thank you very much. I yield back.

Mr. Shimkus. The gentleman's time has expired. The chair recognizes the gentlelady from Colorado, Ms. DeGette, for 5 minutes.

Ms. DeGette. Thank you, Mr. Chairman. I know some of you were worried there weren't a lot of members here, but you had members represent the entire Rocky Mountain west and west coast, so that is pretty darn good, because we are the ones dealing with these issues every day.

I just want – I think Mr. O'Mara is correct, we don't have any silver bullets for solving this problem. You know, being from Colorado, I see this firsthand, and believe you me, we couldn't see the front range for most of August in Denver because of the smoke.

Then I went to Oregon, and the same smoke was in Oregon, and then I went to Vancouver and it was still there for 1 month. This is not normal summer weather for us in the West. The thing we have to realize is there is no one solution. It would be super great if we could just go in and clear out all of this extra wood, and then we wouldn't have as big of a fire risk. Number one, that is not the best management technique for a lot of these areas. But number two, for those of you not from the Rocky Mountain west and west, it is millions and millions of acres that we are talking about. There is no way, even if we had adequate funding, we could go in and clear out this wood.

Secondly, in some of these areas, we really do need to have prescribed burns. We need to have forest management programs that are appropriate for those forests. And I am delighted to see our whole panel sitting here today agreeing with these concepts.

So what can we do? There is a couple of things. Number one, several of you said we have to have adequate funding. And this is such – this is a bipartisan issue for those of us from the west where our colleagues don't seem to understand how important funding is for forest management, no matter what those techniques are.

The second thing is, we have to think about long-term planning. We are not going to be able to solve this air quality issue, or the other related issues, without the long-term planning.

Mr. O'Mara, you talked about the dry soils, the water and everything else from climate change, but there is other issues too. Let's see if they have my picture, if the clerk has my picture to put up. This is a picture that I took in the Pike and San Isabel National forest last month. It is always really fun to go hiking with me, because I stopped and said take a look at this forest. See the trees on the ground? Those trees would not have been on the ground 10 years ago, that is Ponderosa pine, it was all killed by the pine beetle, and they died and they fell down on the forest floor. Then you can see the aspens now that have grown up because of the death of the pine forest. But then, if you look closely you can see the new baby Ponderosa pines growing up.

So this is something the forest has tried to do to naturally recover from the pine beetle infestation. We – in Colorado, we think it is a miracle that all of these millions of acres that look like this have not burned. We have had some devastating fires the last few years, but we did not have devastating fires this year. I don't know why, I think probably luck. But if you want to solve this problem – so these could all be burning.

Now, we all said in Colorado, we need to be able to remove this dead Ponderosa pine, and we did in many areas. But it is millions of acres; it is wilderness areas; it is national forests; it is BLM land. So we have to think of ways where we are going to aggressively address climate change issues, because it is not just the carbon emissions that we are seeing and everything else, it is a whole ecosystem that is impacted.

So I just really want to say, Mr. Chairman, I so appreciate you having this hearing. And I think that there are ways that we can aggressively work in a bipartisan way. But to think we can go down and clear out all the deadwood or just have a few controlled burns, that is not going to solve this problem over this entire massive and beautiful region. Thanks, and I yield back.

Mr. Shimkus. The gentlelady yields back. The chair recognizes the gentleman from Texas, Mr. Flores, for 5 minutes.

Mr. Flores. Thank you, Mr. Chairman. This has been an enlightening panel today. Mr. Boggus, I have a couple of questions for you if we could. We have got something called Good Neighbor Authority, and we have had several people mention that, but nobody's described it. Can you describe Good Neighbor Authority for us?

Mr. Boggus. I guess the easiest answer is, it is a partnership, an agreement we go into with the U.S. Forest Service where they often have either lost the expertise or do not have the personnel available to help them with timber sales, with prescribed burning. And now you have added road building in the latest version into there to help with the management activities on the U.S. Forest Service lands, so we go in partnership with them and help them manage their – the Federal assets, the Federal force.

Mr. Flores. And how does this authority work for the State of Texas? You are the chief forestry officer in the State, how does that work for you?

Mr. Boggus. It is a dialogue that has to go on, and it is something you learn as you go. Like I said, we were an early adaptor, we saw the benefits of this. In Texas, again, we are a private property State, the U.S. Forest Service is only 635,000 acres of forest land in Texas. But that is extremely important because the things that happen on that forest impact the private landowners around the forest. So with insects and disease, with fire and so forth and so, we work with them because we want to help make sure there are other some other programs. Like we had the southern pine beetle prevention program; it is Federal funding through the U.S. Forest Service that we would help with those private landowners get their property thinned and managed around, we kind of call it beetle proofing around the U.S. Forest Service land. We also now, with Good Neighbor Authority, we can work with the U.S. Forest Service partners and get those same – on the inside of the red paint, and get those protected as well and help do some thinning, and keep the forest healthy, that is the whole idea, we want to keep our forests healthy.

Mr. Flores. In your testimony, you mentioned that last year, Texas used prescribed fires on over 200,000 acres. And you also said that burning like this is pretty common across the south. Some States even do high amounts of prescribed burning. What are the challenges that exist with – well, let me rephrase the question. What are the challenges of dealing with prescribed burns versus the challenges of dealing with uncontrolled burns?

Mr. Boggus. A wildfire is much more challenging and much more destructive. Now, a prescribed fire or controlled burn, says what it is, it is prescriptive, you have very set weather parameters, it is lower intensity. So you have less particulate matter, and so what it does is, it fireproofs communities, it fireproofs the area, so it keeps a catastrophic wildfire from happening. It prevents that fire. It is almost like saying fighting fire with fire, because you are making it where the fuel loading is less, you are keeping those four. It is a fire ecosystem in Texas so we are keeping those forests and those lands healthy, and keeping the fuel loading down. So if you were to have a wildfire break out, an uncontrolled, unplanned fire break in through there, it would be much less destructive.

Mr. Flores. And then you also do this adjacent to communities in order to protect those communities from the impact of the wildfire. What do you do to protect the community in the controlled burn process?

Mr. Boggus. Well, obviously, the biggest thing we do, and I guess I will give an example, is our Jones State forest in Texas, which is almost in the city limits of Houston, so it is surrounded by subdivisions. So it is a very difficult place to burn. We have to plan, and part of these things is working with our environmental quality folks, and also working with the community around there, the landowners and homeowners around there, for them to understand if they do have issues, breathing issues, when we are going to it. So there is a lot of communication back and forth that those homeowners and landowners to say here is what is going to happen. If at first, if they are urbanized, urban dwellers, they are not used to seeing smoke. You know, if you didn't grow up in the country and burning your leaves and seeing smoke, it is disturbing. They think it is a wildfire.

So we let them know what is – and we also show them are before and after and the benefits of that fire, the prescribed fire. And now, some of our biggest advocates are the ones that say, Yes, if you have anybody that is against prescribed fire, tell them to call me. So we have a lot of peers that will help and come to our defense, landowners and homeowners.

So you have got to do a lot of outreach with the group, and you have got to do a lot of preparation and planning ahead of time. So the weather has to be right, conditions have to be right so that the smoke cannot be an adverse condition for those homeowners and landowners around the fire.

Mr. Flores. And, of course, one of the ways that the prescribed burns are safer is you do it seasons when you are less likely to have it migrate into an uncontrolled burn.

Mr. Boggus. Absolutely.

Mr. Flores. I am going to try to squeeze in one last question. A controlled burn has an environmental impact, a wildfire has a huge environmental impact. So because a controlled burn has an environmental impact, you have to work with the Texas CEQ on that. Describe that relationship.

Mr. Boggus. That is an ongoing relationship, and that is one of the things we hope to get done, and just started 2 years ago, working with them to look at their rules. We would like to see prescribed fire treated differently than a wildfire, than smoke stacks, than car emissions. It ought to have some lesser because of the good it does and will help in the long-term prevent catastrophic particulate matter getting with a wildfire. So we would like to see the TCEQ treat prescribed burning and those that are done by trained, certified, prescribed fire managers, not just anybody, but that they would have a look at the smoke and emissions from a prescribed fire differently than they do – we are not there yet, but we are having those conversations. And like I said, last week, the chair of the TCEQ said, We need to have more prescribed fire on the ground in Texas, not less. So we are getting there.

Mr. Flores. Thank you. I have a couple of other questions, but I will ask you to respond supplementally to those. We will send those to you. I yield back.

Mr. Shimkus. The gentleman's time has expired. The chair recognizes the gentleman from Georgia, Mr. Carter, for 5 minutes. Mr. Carter. Thank you, Mr. Chairman. And thank all of you for being here. This is certainly a serious problem, particularly in the State of Georgia. Georgia is the number one forestry State in the Nation. As you know we have over 22 million acres of privately owned land, and only about 1.7 million acres of government land. So we are a little bit different from, I think, the scenario that exists out west.

However, we have had our share of fires. We had the West Mims fires, the Okefenokee swamp is in my district. It is truly one of the national treasures of our country. It is a beautiful, beautiful area that I have had an opportunity to visit on numerous occasions.

We had a very serious fire there this past year, the West Mims fire. One of the adjacent property owners to that was telling me about this, and I met with him because he lost a lot of land as a result of the fire that started on the swamp, but spread to his private land. And I will start with you, Mr. Baertschiger, because I see that you worked as a fire training instructor, and a national type 3 incident commander.

I just wanted to ask you, one of the things that was brought to my attention by the private landowner was that they didn't utilize the air support. If they had been able to utilize it quicker, that they could have contained it possibly. Now in all fairness, a swamp fire is a little bit different than other kind of fires, because you have, from what I understand, and I know you all know it a lot better than I do, but the Peat moss, and it is hard to put out, because the water has to rise up, and again, you understand it much better than I do.

But he did make that point that if – and he attributes it to being a problem with the – whether it was low funding and they couldn't afford to utilize the air support, the helicopters that were available. Is there something that you experienced before?

Mr. Baertschiger. Well, there could be – I wasn't there, I don't know what the conditions were. And certain tools work good under certain conditions. If you have a wind blowing in excess of, say, 25 miles per hour, aviation stuff really doesn't help you much. And swamp is tough, because you can't use dozers and other mechanical equipment because they don't go through the swamp very well. So there is challenges with every fire.

But the example you give is very good; it is every forester in this country is exposed to catastrophic wildfire. And our history shows that going back to 1812, but the great Maine and New Brunswick fire, who would have thought that northern Maine and Brunswick would burn up, I think it was 3 million acres, and kill a lot of people.

So, you know, it is hard for me to comment on a fire that I don't have any specifics, but not all the tools work all the time. In Oregon this year, landowners lost 33,000 acres of private timberlands from fires burning off of the Forest Service on to the private lands.

Mr. Carter. Let me ask you, I met with him as I mentioned before, and he owns a lot of forest land in the area in Georgia. And when I met with him, he said a lightning strike is what this originated from. And that generally, the Federal Government will just let it burn out and not even respond to it, is that –

Mr. Baertschiger. You know, it just depends where it is, and, you know, I think – I believe you mentioned it was in a wilderness.

Mr. Carter. Yeah, oh, yeah, in the middle of the swamp, or at least it started, and now it spread on to the private lands.

Mr. Baertschiger. In wilderness comes certain engagement rules, and I think some of that needs to be reviewed.

Mr. Carter. I appreciate that. Let me move to – I wanted to get to Ms. Germann.

Ms. Germann. I am Germann.

Mr. Carter. I wanted to ask you – now you are in Montana, right?

Ms. Germann. Yes.

Mr. Carter. Okay. The practices in Montana, I suspect, are a little bit different than I described in the State of Georgia, particularly in the swamp, and I asked about that in my district. We are not all swamp in Georgia, but in my district we are. I am in south Georgia. But I wanted to ask you about the practice, the forestry practices that you have in Montana. Can you describe those very quickly?

Ms. Germann. Sure. Absolutely. So we have, I will say that 60 percent of the forested land within the State of Montana is managed by the Forest Service. And we have active forest management taking place on State, private and Federal lands. And anything else that you want to –

Mr. Carter. I want to ask you specific about the State information implementation plans, and I guess this is kind of a broad question, and I am out of time, but nevertheless, these have to be approved by the EPA. Is that the way I understand it?

Ms. Germann. Yeah. And I don't have expertise on the State implementation plans. I might ask that my colleague from Idaho – Mr. Carter. I was just wondering if there were any type of barriers that you are having, or any kind of constraints, and how soon did they approve those? How quickly do they approve?

Mr. Shimkus. Quickly, please.

Ms. Anderson. It normally takes an 18-month period for EPA to approve those. So any changes to, like Idaho rules we submit for EPA. It is a very long, drawn-out process.

Mr. Shimkus. The gentleman's time has expired.

Mr. Carter. Thank you very much.

Mr. Shimkus. Chair recognizes the gentleman from California,

Mr. Cardenas, for 5 minutes.

Mr. Cardenas. Thank you very much for having this hearing. Hopefully we can see through the smoke of politics and get things a little more right, than not, in this great country. We have – don't we have some of the best response systems in the world? I mean, aren't we like in the upper tier when it comes to being able to respond to fires and trying to protect life and property? I think everybody pretty much agrees with that. I am not saying we are the best, but we are definitely in the upper tier, right? We have got all that capacity and capability, thank God.

One thing I would like to point out is the wildfires that have been ravaging through California are in excess of anything we have ever seen in the past. For example, 25 years ago, if you had a 4,000-acre fire, that was considered big. Now we have these mega fires that are consuming over 100,000 acres per fire. And then all of a sudden, you have now where people talk about fire season. It is kind of like fire year now, there really isn't a 3- or 4- or 5-month season. Now the situation is so bad, so dire, our forest and our vegetation has dried up so much that the – honest to God truth, as they say, protect yourself and hope and pray that there is not a fire, because there is no season anymore; it could erupt at any given time, and then when it does, we see these mega fires and some of them are raging as we speak.

Another thing as well, I would like to point out this is a responsibility that we need to hopefully get right as policymakers, and as organizations, whether it is local or State or Federal. We need to make sure that we can work together to minimize the negative effects of these devastating fires.

For example, according to the U.S. Forest Service alone, they have spent $2 billion last year just with the fires. That doesn't include the economic loss, et cetera. That is just the Federal investment in that. I truly do believe that we can always do better if we take the opportunity to learn from the past, to learn about what is going on today, to learn about what it is that – how we are going to deal with this issue that many scientists are claiming that some of finest universities, Columbia University, et cetera, are saying that climate change is, in fact, contributing tremendously to some of the fires that are going on today.

I hope we don't argue about the simple fact that we do have a different environment now when it comes to the vegetation, when it comes to the ability to – for Mother Nature to protect itself, and we, as human beings, have to make up the difference. Again, a 4,000-acre fire, not too long ago, was considered big, 100,000 acre fire is now becoming commonplace.

So with that, I would like to also ask the chair and the ranking member coming from California in the future, we can try to glean through the wonderful experts, like the ones we have here today. Maybe we can get somebody from California up here because our disproportionality of being affected by fires as of late is just tremendous.

Again, I don't know if that is a complaint or what have you, I think it is an observation with five members from the California delegation on this subcommittee. Hopefully in the future, we can be a little more –

Mr. Shimkus. Would the gentleman yield?

Mr. Cardenas. Yeah, absolutely.

Mr. Shimkus. You do know the process by which the people are asked are both from the majority and the minority side.

Mr. Cardenas. And that is why I mentioned to both of you, the chair and the ranking member.

Mr. Shimkus. Just wanted to make sure it was clarified.

Mr. Cardenas. But since you brought it up, maybe it is four to one, because we get one person and you get four.

Mr. Shimkus. Will the gentleman yield?

Mr. Cardenas. I will yield.

Mr. Shimkus. These negotiations are always done between the parties, and I see no objection.

Mr. Cardenas. Okay, thank you.

So again, that is why I say it is not so much a complaint, it is just an observation. And hopefully, we can get fortunate enough to have some folks who are dialed in directly within the California scene, especially since it is one of the most dire in the country now when it comes to our fires.

Mr. O'Mara, what will the effect beyond California fire seasons, or as I just called it, fire years, actually if we continue to roll back clean air standards?

Mr. O'Mara. You mean, the challenges that as the fires get worse, the displacing a lot of the air quality benefits that we have accumulated through cleaner power plants, cleaner cars, energy efficiency, all the work that you are doing at State level.

I actually worked for the mayor of San Jose for 3 years and a lot of the work they have been doing – you could undo a lot of that progress unless we deal with the underlying issue: the public health consequences of uncontrolled fires.

Mr. Cardenas. Again, Mother Nature can – if we don't help, can wipe things out, set us back decades, actually.

What holistic steps can Congress, and State and local governments take to do our part in reducing the devastating blazes across California and the U.S.?

Mr. O'Mara. I think we talked a lot about funding today, making sure that we have the resources for the proactive work, through the proactive restoration work. I think there are things we can do to help individuals, make sure there is mitigation money and things like that. But also, making sure we are doing prescribed burns, making sure we are doing good management. And frankly, you have some of the best people in the country in California. The challenge is they don't have the resources they need to do the scale of restoration they need, given the scale of the impacts, and we have to help solve that problem.

Mr. Cardenas. Thank you, Mr. Chairman.

Mr. Shimkus. The gentleman's time has expired.

The chair recognizes the gentleman from the South Carolina, Mr. Duncan, for 5 minutes.

Mr. Duncan. Thank you, Mr. Chairman. Just like hurricanes aren't limited to Florida or the Gulf of Mexico, wildfires are not limited to the west. In 2009, Horry County fire down in Myrtle Beach, same place being affected by Hurricane Florence, experienced 24 miles, 20,000 acres burned, 70 homes destroyed, 2,500 people evacuated. In my district, we have Sumter National Forest, which is 370,442 acres. So national forests and forest fires are not limited to the west.

My wife owns property in Montana. We have been out there since I graduated college in 1989. We have seen what the spotted owl controversy did to the timber industry in the west. I believe that was the beginning of the change of mitigation practices and how forests were managed all throughout the west, not just in Montana. Families that were supported by timber dollars lost their jobs. Ms. Germann from Montana can probably attest a number of saw mills are lost, a number of timber families have been displaced, and the lack of timber activity that you saw in the late 1980s and 1990s; it went away, it went away. And at that point, we started managing our forests differently.

So I traveled to Montana, I was out there this summer in August. I saw all the smoke. I experienced the smell. I saw that all the tourists that came into the Kalispell and Glacier National Airport to go to Glacier National Park, probably didn't see the beautiful scenery of that national park due to the fires, and that was before the Lake McDonald fire. While we were there, had a lightning storm, four lightning strikes, caused four fires, one of them was a Lake McDonald fire. Burned all the way down the lake right there in Glacier National Park. Three of the other lightning strikes from the same storm didn't burn near as much, because they actually hit on property that had been managed properly, and the fires were able to be contained a lot quicker than that in the national park, because we don't do any sort of mitigation efforts in national parks. I am not advocating for that, but I think we ought to at least think outside the box when we are talking about managing fires.

Last summer, not this past August, but a year ago, I was also in Montana, and the Gibraltar Ridge fire, which you are probably aware of up in Eureka, Montana, that was burning very close to our property. So I took it as an opportunity upon myself, and I challenge every Member of Congress and on this committee, to go to a fire camp and visit with the people that are fighting the fires in the fire camp like I did in Eureka, Montana, and then get in the truck with the forest manager, and go out to the fire line and see how these fires are fought. Because I went to the Gibraltar Ridge fire, and I spent 3 hours on the fire line to see the techniques that were being used, mainly mitigation efforts to keep that fire from moving towards where people live, and that personal property to keep it from being destroyed. Other than that, it was just trying to contain the fire, keep more forest acreage from being burned. But they weren't trying to put the fire out at all.

In fact, on in the wilderness study area, there is minimally invasive suppression techniques, missed techniques. So they weren't doing anything up there, but maybe trying to contain it a little bit. Very difficult to get to, I get that.

Having said all of that, we need to back up as a Nation and start talking about how we manage our forests. That means, more timber activity. This is the American taxpayers' resources and it is growing, it is going to regrow. We have practiced timber sales forever. And one of the ways that we can mitigate the pine beetle is cut the timber. She said we don't have a funding stream to do some of these clearing techniques. Guess what? It is called timber sales. They pay for themselves, actually provide revenue back to the government to provide revenue for these expenses.

So let's manage our forests, let's sell some of the timber, and then let's look at shading along roads and near where residential areas are, let's look at fire breaks. Let's look at prescribed burning.

I mentioned the Horry County fire earlier. The reason that fire was so bad and got out of control, and even the firefighters had to employ shelters to let the fire go over and to keep from losing their lives is because the northerners that moved down to South Carolina and occupied in Myrtle Beach, did not like the smoke from prescribed burning. And so prescribed burning didn't happen. And because the prescribed burning did not happen, there was a lot of fuel there. Once that fire started, it burned out of control, because there was so much fuel for it. If we don't manage these fires out west and even in South Carolina with prescribed burning and good management techniques, we are going to see, continue to see, out-of-control wildfires that are very difficult to contain and we are going to pray for a snowfall to put these doggone things out, because that is what they pray for out west is that snow to get there. They see a thunderstorm come in August, that is kind of a double-edged sword. It is providing some moisture to help contain some of that fire, but it is also providing additional lightning strikes.

I was talking to Brian Donner at the Kootenai National Forest Service, a forest ranger there. You may know Brian. He said while they were fighting one fire, a lightning storm came in, they saw lightning hit over on a hill. They saw the tree it hit. They knew right where to go, but before they could get there, because of the amount of fuel that was there, there was 5 or 10 acres already burning and that was very difficult to start containing at that point on the top of that mountain. Had they done prescribed burning and that fuel had gone away, that fire would have been contained a lot quicker.

The last thing I will say, Mr. Chairman, because – Mr. Shimkus. Your time has expired.

Mr. Duncan. – she said in her statement – thank you – over the past century – and this was a good statement by the way, by Ms. Germann – over the past century a cultural fire exclusion unfortunately removed the natural role of fire from the public consciousness, when combined with a reduced level of forest management in many areas of the country, fire exclusion led to the buildup of forest fuels to unprecedented levels. Despite our attempts to manage away wildfire, many our forests are more fire prone than ever. And that is the truth.

And with that, I yield back.

Mr. Shimkus. The gentleman yields back his time. The chair now recognizes the gentleman from California, Mr. Peters for 5 minutes.

Mr. Peters. Thank you, Mr. Chairman. My constituents in San Diego are acutely aware of these issues. I do think it was progress to do the fire fix. I worked really hard on that to make sure that we weren't spending prevention money fighting fires because it just makes it harder to do. You are never going to catch up.

Mr. O'Mara, I have two questions for you, though. Specifically on the Clean Air Act in your testimony, you noted the strange thing where, in terms of calculating your compliance, whether you are in attainment of the National Ambient Air Quality Standards, you are penalized for prescribed burns, but not necessarily for natural burns that happened as a result of not taking care of things. You suggest that EPA can take care of this themselves. Is that not something Congress has to do? Tell me why EPA can change that?

Mr. O'Mara. Yeah, if you go back to the record – the Clean Air Act amendment to 1990, this anthropogenic versus natural kind of distinction isn't as clear-cut as you might think. I mean, it has been an administrative practice, and the challenges that seems to build your prescribed burn and your State implementation plan and basically account for it, a wildfire you have to – it is excluded. You had to get an exemption, because it is kind of considered natural. The challenges – I was in Delaware at the time we were trying to prescribe burns, Delaware has – they have so many challenges being downwind, pollution from coal plants out in the Midwest, there is nowhere to put it in a ship. You have to find a different place in some of those sources to offset. And so it becomes a big burden, so you end up not doing the very thing that would help protect you long-term because of the potential penalty.

Mr. Peters. So you think that that can be addressed at an administrative level?

Mr. O'Mara. I believe so.

Mr. Peters. One other question for you, I like what you did, which was sort of threw out your notes, so I will throw out my notes a little bit and ask you if you were in charge of allocating money for fire, where would you put it first? What would be, you think the highest priority for new fire money?

Mr. O'Mara. There are great collaborative plans that have been on the books for years that don't have the resources to get on the ground. I mean, I think I would prioritize on the interface projects that have the potential of loss of human life. But I would pour money into mitigation, I would pour money into prescribed burns. I would pour money into the collaborative plans that already have buy-in among communities, because they are going to move faster through the process. But we need to move from a couple acres a year to tens of millions of acres a year. We don't have the capacity at this point. I mean, the Forest Service has been, through sequester, their resources were taken down so far in addition to not having the money because of the fire borrowing issue. We have got to rebuild fire capacity in this country at both at Federal and State level.

Mr. Peters. The collaborative plans you are talking about are regional collaborative plans?

Mr. O'Mara. The regional level, yeah.

Mr. Peters. And what sort of management reforms would you like to see enacted, management reforms? I have to confess, I hear a lot of discussion back and forth. It sounds like disagreement, but never quite understand, kind of, what is it that we are fighting over?

Mr. O'Mara. Look, I mean, I think there are places where you could have more efficient processes. There are things where maybe not having to go through the same level of review for individual parts of project if you actually do the analysis at the landscape level. We layered on so many parts of the process.

Mr. Peters. How do I write that down? How do I write that down from here? What does that mean?

Mr. O'Mara. There are ways to do it. I mean, there is some language that Senator Cantwell was working on around Ponderosa pine, basically trying to say Look, if it fits this kind of landscape project, we will have kind of one analysis, one environmental impact review as opposed to having them do every individual discrete component.

So there are some things we can do at the landscape level. Some of that could be done administratively. And if the Forest Service has predictable resources to be able to do that kind of planning, but a lot of these forest plans are 20, 30, and 40 years old. It means we are updating project plans, we are not looking at the landscape level. I think – we would love to work with you on that, because I think that could be bipartisan. I don't think that would be a controversial issue.

Mr. Peters. Obviously, I am particularly interested in the urban forest interface. And I am concerned about the fact that it is not even October and we have already had fire season, we are not even into October. So we are getting ready for what we hear from our local firefighters is as bad a condition or worse than 2003 and 2007, which were the fires that cost San Diego County a lot of property, and money, and damage. So we are very interested in taking you up on that and look forward to talking to you.

Mr. Chairman, I yield back.

The Chairman. [Presiding.] The gentleman yields back. The chair recognizes the gentleman from Ohio, Mr. Johnson. Thanks for joining us. You are recognized for 5 minutes.

Mr. Johnson. Thank you, Mr. Chairman and thanks for holding this hearing today.

You know, while most of our witnesses, many of our witnesses are from western States, these issues are certainly relevant to where I live there in Ohio. I have a significant portion of the Wayne National Forest within my district which will, from time to time, carry out prescribed burns. The Wayne is a patchwork of public and private lands. So these burns are meant to protect human property and reduce potential damages from wildfire, but they can also encourage the growth of plant life, and help ensure oaks, for example, remain prevalent within the forest.

So while we have heard about the benefits of these practices, prescribed burns today, whether that is air quality, safety, et cetera, I would like to discuss the planning that is undertaken before a burn happens. It is crucial that many factors are considered before conducting a burn, such as temperature, humidity, atmosphere stability, wind direction and speed, as well as smoke dispersion.

So a question for either Ms. Germann or Mr. Boggus, or both, along with other resource constraints and other issues, I am sure these factors that I just listed inhibit the ability to accomplish all that is needed to be accomplished over the course of a month or a year. So how do you balance the factors in planning with the need to efficiently manage healthy forests?

Mr. Boggus. You mentioned it already that is planning, you have got to look out. We have a meteorologist on staff because of the very conditions you are talking about. And we have an urbanizing State. I know Montana has 1 million folks, we have 28 million; in Ohio, the same way, a very populated State. You have to take those into consideration. We have 94 percent privately owned. So we don't have the luxury of – if a fire starts, we have got to get on it, and we suppress them all because there are human lives and property, and improved property at stake. And so you have got to plan that. And because of that, you have got to have folks that are dedicated to, we call them predictive services. So they are telling us days and weeks ahead what the weather is going to look like, when is it going to be right.

And so you have these plans written way ahead of time. And you know this is the time, this is the window that this particular piece of land will burn. So then you have Good Neighbor Authority on Federal lands that you work with those, with our partners there. And so, we have got those agreements done well in advance. So you are not like, Oh, my gosh, it is a good day to burn, and you go out and burn. So the planning is crucial.

Mr. Johnson. Sure. Ms. Germann, do you have anything to add?

Ms. Germann. Certainly. I think one of the challenges we were talking about before this hearing is the social license that you have with this. And something that we constantly face, our Federal partners, we as State agencies face when we are planning prescribed burning, the communication piece, so educating the public, getting them to understand the benefits of that.

In the State of Montana, we burn, on average, about 30,000 or 40,000 acres of forested land per year, prescribed burning. We need to do about 10 times that, from an ecological perspective, to really have an impact on fuels reduction. And one of the things that we find the most challenging is getting the public buy in. So I think in addition to all the planning is the communication piece of that that we need to constantly be doing better.

Mr. Johnson. Gotcha. Well, along those same lines, how do you choose what section of forest to address next, particularly if you can't treat every section that needs to be treated? You said you are doing 10, or you are doing 45,000, you need to do 10 times that many. How do you decide which 45,000 acre lot to do?

Ms. Germann. So there is a number of different filters. And I want to clarify that in the State of Montana, we don't just put prescribed burning on the ground, we have to do active mechanical fields reduction before we do that, because our fields are at such unprecedented levels. We use a number of different things statewide, and I will talk about our forest action plan that we are going to be undertaking. What we did do in the State of Montana is our governor did identify 5 million acres of priority treatment, and that was on Forest Service land, under the authority of the 2014 farm bill.

So we match that along with high severity areas, identified by community wildfire protection plans. We use collaborative groups to really help identify where we need to be focusing our treatment. A lot of that is driven by forest pests, insects and disease occurrence, fuel loading, wildfire hazard. We have a lot of that data, and that is where we typically plan our priority treatments.

Mr. Johnson. Thank you, Mr. Chairman. I yield back.

The Chairman. The gentleman yields back. The chair recognizes the gentleman from California, Mr. McNerney, for 5 minutes.

Mr. McNerney. Well, I thank the chairman and ranking member. It feels kind of strange, this morning we were talking about hurricanes, and now we are talking about wildfires. But both of those have some connection to climate impact, so this has to be a holistic discussion.

Now, it seems to me the difficulty is managing forests to prevent and minimize damage, but also protecting health and safety. On the other hand, is it necessary, or will it be necessary at all to prevent – to manage development, so that we don't put people and property at risk at these high risk areas. So my question was sort of a general one for whoever wants to answer: How should we be thinking more holistically about forest fires and management?

Mr. Baertschiger. I would like to respond to that. In your State, which I have been down many times fighting fire, has that Mediterranean climate, and your fuels cure much earlier in the season, and they stayed cured much longer, and then you have the Santa Ana wind event in the southern California. So dispensable space around houses and evacuation routes need to be a lot more thought through because fire in your State burns very quickly. As a firefighter, we say in Oregon, sometimes you can't run fast enough. In California, you can't drive fast enough. So I think that is something you need to take into consideration as you build your communities and expand them into what we call the urban interface, that those conditions are really taken into consideration, defensible space and evacuation routes.

Mr. McNerney. Well, I would like to direct this toward Ms. Germann. How are you working with communities to manage building in these high risk areas?

Ms. Germann. You know, in Montana and some research just came out from one of our groups out of Bozeman that showed that tremendous amount of money is being spent in urban interface in suppressing fires. And I will say, in Montana, we are in the infancy of talking about this from a land use planning perspective. But what we do is DNRC, we are really trying to interface with the local government to help them organize around the tenets of the cohesive strategy. Talk to them about fire-adapted communities, the stuff that we are experts at, at forest management, really helping local governments do that treatment in and around homes; and educate people on the risk of living in the wild land, but urban interface. But from a planning perspective, it is really pretty much in its infancy in the State of Montana.

Mr. McNerney. So do you feel the local communities are responsive to your advice and input?

Ms. Germann. Certainly, absolutely. We pride ourselves in really excellent relationships with local governments. We have a local government forest adviser who is engaging with county commissioners and volunteer Fire Departments on engaging with the Forest Service, which is the predominant landowner, forest landowner around the communities about suppression efforts, about forest fuels reduction, and certainly, we help deliver a lot of that education to private landowners within our communities.

Mr. McNerney. Mr. O'Mara, is there a lack of funding that we can address at the Federal level to improve how we as a Nation handle wildfire management?

Mr. O'Mara. Yes. I mean, I think it is amazing what the Congress did in the last session, fixing the fire borrowing practice; it is still an underinvestment. I can say all Americans are Libertarians until they need help. We have to figure out a way to monetize some of these costs. They are putting people in harm's way, they are putting firefighters in harm's way. It is the same thing in flood insurance, it is the same thing. We are basically paying people to be on the – be in more risky areas. I just think – I think we are billions of dollars short in terms of the amount of money that is used toward active restoration annually, that is the kind of level of funding that we are going to need, because Chairman McKinley and I have gone back and forth on many issues. He is exactly right. I want to say when he is not here. Because we are not talking the east coast forest enough. The east coast forests and the Great Lakes forests have equal threats, they are just a couple of years behind in terms of the temperature patterns.

Mr. McNerney. Well, I think one of the big controversies or areas of disagreement is whether we should use suppression or management. From the science that I have seen, the fires can be managed better, and it gives the forest a better chance to recuperate and create national fire breaks and natural water sheds and so on. So I wouldn't rush to one or the other. But I would lean toward management, in my opinion. Thank you, I yield back.

The Chairman. The gentleman yields back, I want to thank our panelists for being here, we will send Mr. McKinley a video of your comments where you agree with him. I don't know how that is going to play out. But we do appreciate it. Our work is better informed by your participation, I know some of you, including the Senator, have traveled great distances, and we thank you for doing that.

Seeing there are no further members to ask questions for the first panel, I would like to thank all of our witness for being here today. Before we conclude I would like to ask unanimous consent to submit the following documents for the record: Two academic reports entitled Prescribed Fire in North American Forest and Woodlands, and Prescribed Fire Policy, Barriers, and Opportunities; and document from the National Academy of Sciences, called, The Impact of Anthropogenic Climate Change on Wildfire Across the Western U.S. Forests; an article from GeoHealth, Future Fire Impacts on Smoke Concentrations, Visibility and Health in the Contiguous United States; Washington Post editorial board, We Won't Stop California's Wildfires if We Don't Talk About Climate Change; New York Times article, Trump Inaccurately Claims California is Wasting Waters as Fires Burn. The Scientific American article Fuels by Climate Change Wildfires Erode Air Quality Gains; and a document from the National Wildlife Federation, Mega Fires.

And in pursuant to committee rules, I remind members they have 10 business days to submit additional questions for the record. I ask that our witnesses respond to those questions within 10 business days upon receipt of those questions. And so again, thank you all for participating in this very important hearing, and without objection, this subcommittee is adjourned.

[The information follows:]

[Whereupon, at 3:15 p.m., the subcommittee was adjourned.]

PRESCRIBED FIRE IN NORTH AMERICAN FORESTS AND WOODLANDS: HISTORY, CURRENT PRACTICE, AND CHALLENGES

Kevin C. Ryan[*], Eric E. Knapp[†] and J. Morgan Varner[‡]

> Whether ignited by lightning or by Native Americans, fire once shaped many North American ecosystems. Euro–American settlement and 20th-century fire suppression practices drastically altered historic fire regimes, leading to excessive fuel accumulation and uncharacteristically severe wildfires in some areas and diminished flammability resulting from shifts to more fire-sensitive forest species in others. Prescribed fire is a valuable tool for fuel management and ecosystem restoration, but the practice is fraught with controversy and uncertainty. Here, we summarize fire use in the forests and woodlands of North America and the current state of the practice, and explore challenges associated with the use of prescribed fire. Although new scientific knowledge has reduced barriers to prescribed burning, societal aversion to risk often trumps known, long-term ecological benefits. Broader implementation of prescribed burning and strategic management of wildfires in fire-dependent ecosystems will require improved integration of science, policy, and management, and greater societal acceptance through education and public involvement in land-management issues.

Front Ecol Environ 2013; 11 (Online Issue 1): e15–e24, doi:10.1890/120329.

Wildland fire has impacted most landscapes of the Americas, leaving evidence of its passing in the biota, soils, fossils, and cultural artifacts (Swetnam and Betancourt 1990; Delcourt and Delcourt 1997; Platt 1999; Ryan et al. 2012). Many terrestrial ecosystems reflect this long evolutionary history with fire and require periodic fire to maintain species composition

---

[*] USDA Forest Service Rocky Mountain Research Station, Missoula, MT (retired) (kcryan@fs.fed.us).
[†] USDA Forest Service, Pacific Southwest Research Station, Redding, CA.
[‡] Mississippi State University, Department of Forestry, Forest and Wildlife Research Center, Mississippi State, MS.

and stand structure and function (Abrams 1992; Agee 1993; Pausas and Keeley 2009).

In a Nutshell:

- Industrial-era land-use changes and fire exclusion have greatly modified fire regimes across much of North America, and the ecological consequences of these policies are becoming better understood
- Increased use of prescribed fire and ecologically beneficial management of wildfires will be necessary to treat fuels and restore fire-adapted landscapes
- Restoration of the multi-scale structural complexity that was historically produced by fire will benefit from a variable fire regime, including burns at different times of the year, under different weather and fuel-moisture conditions, and the use of heterogeneous ignition patterns
- While science has and continues to play a vital role in fire management, sociopolitical constraints – including public acceptance, aversion to risk, and inadequate funding – are often greater barriers to the use of fire than remaining ecological unknowns

---

The presence of fuels and a source of ignition are necessary for wildland fires to occur. Variations in fire spread and intensity across landscapes are dependent on the physical and chemical characteristics of these fuels, with fuel moisture and fuelbed continuity being two of the most important factors. An abundance of fine (high surface area-to-volume ratio), dry fuels that are continuous or interconnected is required for fire to spread. Cold- or moisture-limited ecosystems are often fuel-limited because combustible biomass accumulates slowly and the continuity of the fuelbed takes longer to redevelop following a fire. Wet forests develop fuelbed continuity more quickly but may also be effectively fuel-limited because the fine fuels are rarely dry enough to burn.

Intermediate to these extremes are a range of ecosystems that produce abundant fine fuel and are seasonally dry and susceptible to ignition from lightning or humans. Rates of fuel accumulation and prevalence of ignition sources varies by region and ecosystem across North America (Knapp et al. 2009). Within regions, fire potential also varies year to year, under the influence of global circulation patterns such as the El Niño–Southern Oscillation (ENSO; Swetnam and Betancourt 1990; Ryan et al. 2012). The southeastern US coastal plains and southwestern mountain ranges experience frequent lightning storms; when lightning strikes dry fuels, for example, in the days prior to summer monsoon rains (Figure 1; Flagstaff, Arizona and Ocala, Florida), numerous fires result (Swetnam and Betancourt 1990; Stambaugh et al. 2011).

Major conflagrations commonly occur during La Niña episodes, when monsoonal rains are delayed or weak. These areas recover fuel continuity quickly and are characterized by high fire frequency. In contrast, soaking summer rains hamper lightning ignitions in the deciduous hardwood forests of northeastern North America (Figure 1; Athens, Ohio). In this region, fuels are combustible mainly during autumn–spring dormancy, the period when sunlight can dry the newly-fallen leaf litter. Lightning is rare during this time and fires are therefore primarily human-caused (Schroeder and Buck 1970; Guyette and Spetich 2003). Lightning fires are largely restricted to ridges and sandy plains that favor the development of more open pine–oak (*Pinus* and *Quercus* spp, respectively) forests, and where more rapid drying of surface fuels is possible (Motzkin et al. 1999; Keeley et al. 2009). Much of western North America is typified by an extended summer dry season (e.g., Figure 1; Yosemite National Park, California). "Dry" thunderstorms – those that lack wetting rains – are a major source of summer fires in the western mountains, particularly during droughts. Lightning is also the dominant source of large, landscape-scale fires in the boreal forests of Alaska and northern Canada (Krezek-Hanes et al. 2011). In many areas of North America, relatively recent settlement of rural woodlands is shifting the proportion of human versus lightning ignitions (Peters et al. 2013).

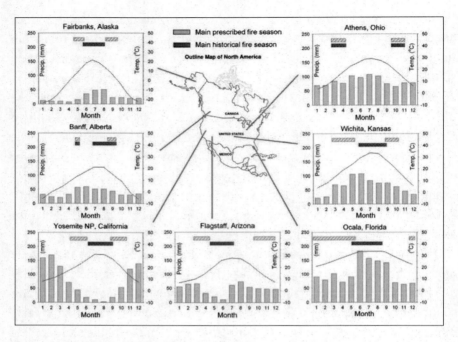

Figure 1. Climographs consisting of monthly average temperature (blue line) and precipitation (grey bar), and the approximate time of year of the peak historical and prescribed fire seasons from seven representative areas in North America with active prescribed fire programs. Cyclic patterns associated with general circulation patterns (e.g., ENSO) may expand the fire season in a given year and occasional large fires occur under extreme meteorological events.

Humans and Fire Prior to Euro–American Settlement

Humans migrated to the Western Hemisphere at least 14 000 years before present (Goebel et al. 2008) and used fire for heat, light, food preparation, and hunting (cf Nowacki et al. 2012; Ryan et al. 2012), but the degree to which human-caused fires were agents of land-cover change is unknown because of the spatial and temporal limitations of paleological data. Questions therefore remain about the extent to which pre-Columbian fires were of natural or human origin (Boyd 1999; Vale 2002). In areas of high lightning density, such as in the mountains of the US Southeast and Southwest, fire frequency was most likely limited by the recovery rate of fine fuels. In Pacific Coast forests and in the temperate deciduous forest biome of eastern North America, the rarity of dry-season lightning suggests that humans were a major ignition source (McClain and Elzinga 1994; Brown and Hebda 2002; Kay 2007; Abrams and Nowacki 2008); while lightning fires occur in these systems, it is difficult to explain the frequency of historic burning without human ignitions (Keeley 2002; Guyette and Spetich 2003; Spetich et al. 2011).

Native Americans used fire for diverse purposes, ranging from cultivation of plants for food, medicine, and basketry to the extensive modification of landscapes for game management or travel (Pyne 1982; Anderson 2005; Abrams and Nowacki 2008). Although landscape-scale fire use ended with nomadic hunting practices, the smaller scale use of fire to promote various plant materials remains an integral component of traditional ecological knowledge in American Indian cultures (Anderson 2005).

An estimated 21 million indigenous people inhabited North America at the time of initial European settlement (Denevan 1992). Eurasian diseases transmitted by these early settlers decimated native populations. Many regions show a marked reduction in fire frequency at the same time as this population decline (Spetich et al. 2011; Power et al. 2012). This period also coincides with the cold, wet Little Ice Age climate anomaly (Power et al. 2012), which may also have played a role in reducing the number of fires. For these reasons, by the time substantial European immigration began in the 17th century, settlers encountered landscapes that were adjusting to less frequent burning.

Humans and Fire after Euro–American Settlement

European settlers caused major changes in fire regimes throughout North American forests. Logging was associated with land clearing for agriculture, as well as providing fuel for heating, powering steam engines, and industrial production. Unregulated forest harvesting during the 19th and early 20th centuries generated logging slash (residual coarse and fine woody debris) that contributed to catastrophic wildfires (Haines and Sando 1969;

Pyne et al. 1996). In the US, the societal and legal responses to these fires made wildland fire suppression a dominant activity in federal, state, and private forest management (e.g., USFS 10 AM Policy of 1935). Fire factored into the creation of several federal land-management agencies (e.g., US Forest Service [1905], US National Park Service [1916], and the US Bureau of Land Management [1946]) and similar forest conservation agencies at the state level (Pyne 1982). Without exception, agency policies coupled with propaganda on the benefits of fire prevention (e.g., Smokey Bear) were designed to control the impacts of fire through active fire prevention and suppression (Pyne 1982).

Early organized efforts at fire control by fledgling government agencies were hampered by the lack of roads and fire suppression infrastructure. Airplanes and equipment freed up by the end of World War II, as well as intensified road building for logging to support post-war housing demand, helped to bring effective fire suppression to all but the most remote areas, such as the northern boreal forests.

The combination of fire suppression and the decrease in burning by Native Americans dramatically altered the fire regime across much of North America. The eastern US experienced a steep decline in fire occurrence (Nowacki and Abrams 2008). In the western US, the total area burned declined sharply for some decades, reaching its minimum during the 1970s (Agee 1993; Leenhouts 1998). Since then, the trend has been toward increasing wildfire activity (Westerling et al. 2006; Littell et al. 2009), despite extensive suppression efforts. In Canada, yearly burned area increased from 1959 to the 1990s, then declined somewhat, except in the western provinces (Krezek-Hanes et al. 2011). Regardless of regional differences, the land area being burned today across much of North America is far less than what was burned historically. Leenhouts (1998) estimated that in the conterminous US, burning in the late 20th century was 7–12 times less prevalent than in pre-industrial times. In California, Stephens et al. (2007) estimated that 18 times less area was burned annually between 1950 and 1999 than had burned prior to that time. A compilation of studies of Canadian boreal forests indicated an average modern burn rate approximately five times less than the historical burn rate (Bergeron et al.

2004). Similar statistics for Mexico and Central America are not as well developed; here, fires continue to burn across large areas in some years, and ecosystems vary between experiencing less than and more than historic levels of fire (Rodríguez-Trejo and Fulé 2003; Martínez Domínguez and Rodríguez-Trejo 2008).

Ecological Consequences of Fire Exclusion

Excluding fire from previously fire-frequent ecosystems results in major changes in ecosystem structure, composition, and function across a variety of scales (Covington and Moore 1994; Keane et al. 2002; Varner et al. 2005). The consequences of suppression-altered fire regimes include a reduction in or loss of ecosystem services, and vastly altered fuels and potential future fire behavior. Without the disturbance of periodic fire, tree density increases (Figure 2) and landscape structure homogenizes (Taylor 2004; Hutchinson et al. 2008; Nowacki and Abrams 2008). The influx of fire-sensitive species alters community composition, stand structure, and ecosystem processes (Keane et al. 2002; Rodewald and Abrams 2002; McShea et al. 2007; Alexander and Arthur 2010; Maynard and Brewer 2012). Canopy infilling by shade-tolerant, fire-sensitive trees and accumulated litter in unburned forest floors can lead to reduced cover and diversity (Hiers et al. 2007; Engber et al. 2011). Plant species that benefit from disturbance and exposed bare soil typically decline (Harvey et al. 1980; Gilliam and Platt 1999; Knapp et al. 2007). The effects of fire exclusion also affect animal communities. Loss of herbaceous species in long-unburned forests has been associated with reduced butterfly diversity compared to more recently burned forests (Huntzinger 2003). In southeastern pine savannas and woodlands, avian, herpetofauna, and mammalian diversity have declined substantially. The rarity of many endangered wildlife species, including the red-cockaded woodpecker (*Picoides borealis*) and gopher tortoise (*Gopherus polyphemus*), is thought to be largely due to the alteration of habitat caused by the lack of fire (Means 2006).

Figure 2. Ponderosa pine (Pinus ponderosa) forest at the Fort Valley Experimental Forest near Flagstaff, Arizona, showing: (a) effects of fire exclusion and (b) adjacent stand after multiple prescribed burns. In the absence of fire, forests throughout the southwestern US have become dense with young trees that not only make prescribed fire more difficult to implement but also contribute to uncharacteristically intense wildfires.

In drier portions of western North America, greater surface fuel continuity in combination with the influx of conifer seedlings and saplings contributes to higher fire intensity and severity, and an increased probability of crown fires (Agee and Skinner 2005). In contrast, fire exclusion in fire-prone landscapes of eastern North America (particularly oak, southern pine, and oak–pine ecosystems), is associated with the invasion of fire-sensitive species with less flammable litter, more shaded and moister microclimatic conditions, and reduced fire activity. The result is a positive feedback cycle, termed "mesophication" by Nowacki and Abrams (2008), with lower potential for burning reinforcing the advantage for the invading shade-tolerant, fire-sensitive species.

Restoring Fire as a Landscape Process

In North America, recognition of the ecological benefits of prescribed burning was slow in coming and varied geographically. Fuel accumulation and loss of upland game habitat occurred especially quickly in productive southern pine forests and woodlands and ecologists in the southeastern US promoted the use of fire in land management from early on (e.g., Stoddard

1931; Chapman 1932). In spite of their convincing arguments, fire in the southeastern US (and elsewhere) was still frequently viewed as incompatible with timber production due to the potential for injury to mature trees and the inevitable loss of tree seedlings. Since then, research in numerous ecosystems has helped shape greater public recognition of fire's integral role in maintaining "fire-dependent" plant communities. However, contemporary fires fueled by biomass that accumulated in the absence of fire now pose a greater risk of damage to private property, public infrastructure, and ecosystems. Numerous studies have documented the capacity for prescribed burning to mitigate extreme wild-fire behavior and uncharacteristically severe fire effects (Agee and Skinner 2005; Finney et al. 2005; Prichard et al. 2010; Cochrane et al. 2012), further reinforcing the importance of fire management (Ryan and Opperman 2013). Nevertheless, the tension between risks and recognized benefits remains.

The extent to which fire has been incorporated into management protocols varies across regions. In the US, approximately one million ha are burned annually as a result of prescribed fire (NIFC 2013a). Between 1998 and 2008, US federal agencies also actively managed an average of 327 lightning-caused wildfires for the purpose of restoration, and these burned an additional 75 000 ha annually (NIFC 2013b). US federal fire managers still have latitude to allow some lightning fires to burn to provide resource benefits, but since a 2009 policy change, hectares treated in this way are no longer counted separately from total wildfire hectares. In Canada, a small percentage of wildfires in remote areas are allowed to burn or are not aggressively suppressed; these account for the majority of acres burned (Taylor 1998).

Parks Canada and some First Nations conduct prescribed burns on a limited basis (Weber and Taylor 1992), but landscape-scale prescribed burning for ecosystem restoration is still relatively uncommon (Taylor 1998). While statistics for Mexico and Central America indicate a preponderance of human-caused fires, most are either escaped agricultural and pastoral burns or intentional burns that lack clear ecological objectives (Rodríguez-Trejo and Fulé 2003; Rodríguez-Trejo 2008). Despite successes in the development of robust prescribed burning programs, especially in the southeastern US (Stephens 2005), almost nowhere has the use of fire kept pace with or even approached historic levels (Leenhouts 1998; Stephens et al. 2007). The reasons for this "fire deficit" are numerous and can be attributed to lingering questions about the comparability of prescribed or managed burning to pre-industrial fire, as well as legal, political, and operational challenges that accompany burning in the modern era.

Is Prescribed Fire an Ecological Surrogate for Historical Fire?

Where restoration or maintenance of ecological processes is the goal, questions persist about how well current prescribed fires emulate the ecological effects of pre-suppression era fires. One major area of concern is the extent to which current fuel loading exceeds pre-industrial levels. Many fire effects are closely tied to the amount of fuel consumed (Ryan 2002; Knapp et al. 2007, 2009), and initial restoration burns after long fire-free periods can therefore lead to undesirable effects (Ryan and Frandsen 1991), such as killing or stressing large remnant trees, including those of normally very fire-resistant species (Figure 3; Ryan and Reinhardt 1988; Varner et al. 2005; Hood 2010; Harrington 2012).

Variability in fuel distribution generated by periodic fire caused historical fires to burn in a patchy mosaic (e.g., Show and Kotok 1924). This created numerous unburned refugia where fire-sensitive plant species or small non-mobile animals survived to recolonize burned areas. Increased forest density and accumulation of litter, duff, and wood debris has produced a more continuous, uniformly flammable fuelbed (Knapp and Keeley 2006). As a result, in long-unburned areas, prescribed fire or wildfire often leave few such refugia.

Subsequent fires at shorter intervals can re-establish patchiness (Figure 4). However, prescribed fires are also often ignited in linear strips or at multiple points along regular grids (Figure 5a). Uniform ignition, driven by the operational need to maintain control, produces more uniform burns with fewer residual unburned patches.

In contrast, wildfires typically ignite landscapes in large fingered fronts or via lofted embers (spotting), both of which lead to substantial heterogeneity in burn patterns. Our understanding of how refugia and heterogeneity affect organisms at different spatial scales remains incomplete (Knight and Holt 2005; Collins et al. 2009).

Figure 3. Reintroduced fires in this longleaf pine (Pinus palustris) forest in northern Florida ignited accumulated fuels on the forest floor (a, b) that mound adjacent to the tree bole (arrow in [c]). Burning of accumulated fuels can stress and kill large trees in these ecosystems and many other fire-excluded North American forests.

Many prescribed burns are conducted in different seasons and under higher moisture conditions than historical fires (Figure 1; Knapp et al. 2009). A common criticism is that such "cool season" burns fail to achieve fuel consumption and restoration goals.

In the western US, the lack of fire crew availability frequently pushes prescribed burning to the cool spring or fall margins of the fire season, whereas the majority of the area historically burned in the summer, when conditions were warmer and drier (Figure 1). In the southeastern US, dormant-season burns are often preferred over late spring/summer (i.e., lightning-season) burns (Figure 1) to moderate effects, reduce the probability of fire escape, and avoid impacts on breeding birds. Such dormant-season burns are generally less effective for killing encroaching fire-sensitive hardwoods (Streng et al. 1993).

In western woodlands and montane forests, fires historically maintained low tree density by thinning primarily susceptible juveniles (Cooper 1960; Kilgore 1973), but after prolonged fire exclusion many invading trees become large and thickbarked enough to resist stem injury from low-intensity fires (Schwilk et al. 2009; Engber et al. 2011). Prescribed fire alone, especially at the low end of the intensity spectrum, is therefore often inadequate for meeting forest restoration and management goals, and may require augmentation by mechanical means. In other situations, excess fuels, especially around the base of large pines (Figure 3), may lead to excessive stem and root injury and death of the remnant trees that managers most wish to protect (Varner et al. 2005; Hood 2010).

Variations in fire susceptibility among organisms as a result of differing phenology or life-history stage at the time of burning can lead to species shifts (Kauffman and Martin 1990; Howe 1994). However, the majority of studies show little or no influence of timing of burns, relative to other factors such as fire intensity, that also typically vary with season (Knapp et al. 2009). Over the long term, many plant and animal populations appear to be most strongly influenced by how fire alters their habitat, regardless of burning season (Knapp et al. 2009).

Figure 4. Unburned patch resulting from reduced flammability of prostrate ceanothus (Ceanothus prostratus) within a prescribed burn in heterogeneous fuels, 10 years after the first prescribed burn in Klamath National Forest, California. Such potential fire refugia may play an important role in the resilience of species to wildfire or prescribed fire, and are less common in long-unburned areas.

The restoration of structural complexity that was historically generated by frequent low- to mixed-severity wild-fire is a key goal of current federal forest land management. When prescribed fire is used, restoration benefits from a variable fire regime – burning at different times of the year, under different weather and fuel moisture conditions, and employing variable ignition patterns (Knapp et al. 2009), all factors that complicate fire management operations. With prescribed burning, maintaining control of the fire is a primary concern, thereby encouraging the use of low-intensity fire. In addition, common ignition patterns, such as strip head fires (linear strips of fire ignited evenly and in close succession at right angles to the slope and/or wind direction; Figure 5a), are designed to homogenize fire behavior, which in turn also tends to homogenize fire effects. Greater randomness in ignition, including variable, ground-based firing patterns (Figure 5b) or aerial ignition, may increase heterogeneity and better emulate the complexity that historical burning once produced. Since forest management has embraced stand- to landscape-scale structural complexity as a tenet, prescribed fire objectives should ideally seek to incorporate these same outcomes (Noss et al. 2006). Strategic management of wildfires is an especially promising means of generating heterogeneity, due to the inherent variation in fire intensity and severity within wildfire boundaries (Collins et al. 2009). In addition, strategic management of wildfires may allow larger land areas to be burned than can be realistically treated with prescribed fire.

Legal, Political, and Operational Challenges in a Risky World

Research has improved our understanding of the ecology associated with prescribed burning and will continue to play an important role in successful fire management. However, ecological concerns typically pale in comparison to legal, political, and operational challenges. In the US, tension exists between fire and a variety of socioenvironmental values. Prescribed fire treatments must be conducted within the framework of a suite of environmental laws, including the National Environmental Policy Act, the Clean Air Act, the Clean Water Act, and the Endangered Species Act, and the resulting analysis and review processes that accompany land management often lead to conflicts. For example, while the Clean Air Act had the beneficial effect of reducing hazardous particulates from industry and automobiles, it has also made the use of prescribed fire or allowing wildfires to burn much more difficult. Smoke was likely an ever-present reality of fire seasons in the pre-Euro–American landscape (Leenhouts 1998; Stephens et al. 2007), but decades of increasingly effective fire suppression and urbanization has resulted in a public that is out of touch with landscape burning. Recent transmigrations have fragmented the land with subdivisions (Gude et al. 2013; Peters et al. 2013) and many people are unaware of the past prevalence of fire and smoke. Prescribed fire is a point pollution source and therefore easy to regulate. In times of poor air quality, it is often politically less challenging to limit land managers' fire use than to constrain other sources of pollution (e.g., emissions from automobiles or industry).

While some environmental laws have bolstered the case for managers to use fire (e.g., the federally listed fire-obligate red-cockaded woodpecker and many others; Means 2006), in other situations, environmental laws can actually impede prescribed burning (Quinn-Davidson and Varner 2012). The Endangered Species Act requires managers to analyze the immediate short-term risks associated with actions such as prescribed burning, but not the long-term risks associated with inaction. Thus, the law creates a disincentive to treat lands inhabited by endangered species. Short-term risks to a species (e.g., displacement, injury, direct mortality) should ideally be balanced against long-term habitat needs. For example, in western forests, fire may consume snags used for nesting by the northern spotted owl (*Strix occidentalis caurina*), a species officially listed as "threatened" in the US and "endangered" in Canada, but fire also creates snags in the long term, and Irwin et al. (2004) hypothesized that spotted owls abandon nest sites due to reduced foraging efficiency in areas where forest density has increased in the absence of fire. In addition, when wildfire occurs after long periods of exclusion, it can burn at a higher intensity and cause nest sites and surrounding forest habitat to be lost for decades or centuries (e.g., North et al. 2010). Similar conflicts between short- and long-term risks have been described for the effects of fire on endangered bat species in the hardwood forests of central North America (Dickinson et al. 2009), where heat and smoke may be disruptive in the short term but will potentially have positive effects on snag production, canopy openness, and prey availability over the long term.

Beyond the ecological considerations are two additional sources of tension: public acceptance and adequate funding (Quinn-Davidson and Varner 2012). Throughout North America, there are wide variations in the public's willingness to accept smoke, visual impacts, and increased short-term risks associated with prescribed burning (Weber and Taylor 1992; McCaffery 2006). The disparity in the type of land ownership and differences in the legal, political, and cultural environments affect the attitudes of fire managers and communities in these fire-prone regions (McCaffrey 2006; Quinn-Davidson and Varner 2012). Wildlands in the southeastern US are predominantly privately owned, whereas wildlands in the western states are mostly public. In several southeastern US states, prescribed burning is widely considered a public "right." Legislation protects burners, whether government or private, unless thresholds of negligence have been exceeded (Yoder et al. 2004; Sun and Tolver 2012). Florida has long stood as the model for prescribed burning legislation (e.g., Wade and Brenner 1992), and is emulated by other southeastern states (Sun and Tolver 2012). Further testament to the importance of prescribed burning in the Southeast are the long-standing Prescribed Fire Councils that originated in Florida and that have since expanded to other fire-prone southeastern states. These "communities of practice" (Wenger 2000) have been influential in the legislative process and in the training and education of managers and land owners. In contrast, fledgling Prescribed Fire Councils in the western US have yet to petition for protective legislation for burners.

Figure 5. Prescribed fire ignition patterns in Klamath National Forest, California. Ignition patterns can influence fire effects. Some common patterns include: (a) strip head fire, with evenly spaced strips placed sequentially from higher to lower elevations within the unit; and (b) tree-centered spot firing, with the objective of minimizing flame lengths under desired trees and producing variable flame lengths elsewhere.

Prescribed burning can be negatively affected by those rare mistakes or unexpected events that can overwhelm understanding of their ecological and economic benefits. Over 99% of prescribed fires are successfully held within planned perimeters (Dether and Black 2006). When prescribed burns go well, the immediate effects are often little noticed and landscape changes are gradual. But when burns escape, the consequences for future burning can be enormous. For example, high winds caused the May 2000 Cerro Grande prescribed fire in New Mexico's Bandelier National Monument to breach control lines and burn about 19 000 ha and over 250 homes. In Colorado, during the spring of 2012, embers from a seemingly extinguished 4-day-old prescribed burn reignited in high winds, resulting in the Little North Fork Fire that killed three people and destroyed 27 homes. Such high-profile events have the immediate effect of halting prescribed burning until fact-finding concludes; more importantly, they fuel public fear and increase skepticism regarding prescribed burning.

Managers often receive public praise for suppressing wildfires but receive little recognition when conducting successful prescribed burns or allowing wildfires to burn for resource benefits. Disincentives for using fire, as well as societal intolerance of risk and a tendency toward short-term planning, lead to a focus on minimizing short-term risks (i.e., injury to species from heat or smoke, fire escape). Long-term risks (and ecological consequences) posed by fire exclusion attract less discussion and decision-making attention than they probably should.

The risk of escape is greater when weather and fuel moisture conditions approximate historical burning conditions. Prescriptions are therefore often conservative, requiring fuel moisture, relative humidity, and wind speeds that minimize the chance of fire escape. Unfortunately, such conditions are uncommon, resulting in narrow burn windows of only a few days per year in many western landscapes (Quinn-Davidson and Varner 2012). Infrequent favorable conditions increase competition for resources and air quality permits, which are often major hindrances to burning. Thus, sociopolitical factors rather than ecological rationales often drive decisions regarding when and where treatments occur.

Conclusion

Anthropogenic and lightning fires shaped North American landscapes for millennia, so that many ecosystems are dependent on periodic fire to maintain important components (Abrams 1992; McClain and Elzinga 1994; Delcourt and Delcourt 1997; Pausas and Keeley 2009; Nowacki et al. 2012).

There is, however, still much to be learned, particularly with respect to how fire regimes (i.e., the frequency, timing, and severity of fire) affect stand-level processes, and how fire relationships change at increasing temporal and spatial scales. Most studies are relatively short term and often use data collected from small plots, whereas fire management planning occurs across decades and over large landscapes (Keeley et al. 2009).

Technology has greatly expanded our ability to modify fire regimes through fire suppression, prescribed burning, and mechanical manipulation. The ecological legacy of past practices has altered systems, in some cases irrevocably. Future climate conditions will further confound our understanding, and the magnitude and scale of accompanying changes to vegetation and fuels may limit our capacity to respond. These uncertainties constrain our ability to reintroduce fire to accomplish a suite of societal benefits, including protecting lives and property, enhancing ecosystem services, ecological restoration, and biological conservation. Experience indicates that neither laissez faire fire management nor full suppression will accomplish these goals. With current limits to prescribed burning, many managers have turned to mechanical surrogates (e.g., thinning and pile burning). Allowing lightning-ignited wildfires to burn for resource benefits where consistent with local management plans offers promise for restoring large, relatively roadless landscapes (Noss et al. 2006; Collins et al. 2009) but may be impractical in more developed areas.

Humans have been, and will continue to be, a dominant force in shaping the landscape (Denevan 1992; Nowacki et al. 2012; Ryan and Opperman 2013).

Prescribed burning and managed wildfire have been, and should continue to be, major tools for affecting that process. The challenge for all natural resource management centers around not only conserving the species but also preserving and/or restoring biophysical processes. Given the current lack of public awareness and social acceptance (McCaffrey et al. 2013), subdivided and fragmented landscapes (Gude et al. 2013; Peters et al. 2013), and limited funding, expansion of prescribed fire programs will entail a redoubled effort to integrate fire and ecological sciences into management and policy.

Acknowledgments

We wish to thank our colleagues for conversations over the years that helped shape our thinking about the role of prescribed fire in the past, present, and future. Comments by R Keane substantially improved the manuscript.

References

Abrams MD. 1992. Fire and the development of oak forests. *BioScience* 42: 346–53.
Abrams MD and Nowacki GJ. 2008. Native Americans as active and passive promoters of mast and fruit trees in the eastern USA. *Holocene* 18: 1123–37.
Agee JK. 1993. Fire ecology of Pacific Northwest forests. Washington, DC: Island Press.
Agee JK and Skinner CN. 2005. Basic principles of forest fuel reduction treatments. *Forest Ecol Manag* 211: 83–96.
Alexander HD and Arthur MA. 2010. Implications of a predicted shift from upland oaks to red maple on forest hydrology and nutrient availability. *Can J For Res* 40: 716–26.
Anderson MK. 2005. Tending the wild: Native American knowledge and the management of California's natural resources. Berkeley, CA: University of California Press.
Bergeron Y, Flannigan M, Gauthier S, et al. 2004. Past, current and future fire frequency in the Canadian boreal forest: implications for sustainable forest management. *AMBIO* 33: 356–60.
Boyd R. 1999. Indians, fire, and the land in the Pacific Northwest. Corvallis, OR: Oregon State University Press.
Brown KJ and Hebda RJ. 2002. Ancient fires on southern Vancouver Island, British Columbia, Canada: a change in causal mechanisms at about 2000 ybp. *Environmental Archaeology* 7: 1–12.
Chapman HH. 1932. Is the longleaf type a climax? *Ecology* 13: 328–34.

Cochrane MA, Moran CJ, Wimberly MC, et al. 2012. Estimation of wildfire size and risk changes due to fuels treatments. *Int J Wildland Fire* 21: 357–67.

Collins BM, Miller JD, Thode AE, et al. 2009. Interactions among wildland fires in a long-established Sierra Nevada natural fire area. *Ecosystems* 12: 114–28.

Cooper CF. 1960. Changes in vegetation, structure, and growth of southwestern pine forests since white settlement. *Ecol Monogr* 30: 129–64.

Covington WW and Moore MM. 1994. Southwestern ponderosa forest structure: changes since Euro–American settlement. *J Forest* 92: 39–47.

Delcourt HR and Delcourt PA. 1997. Pre-Columbian Native American use of fire on southern Appalachian landscapes. *Conserv Biol* 11: 1010–14.

Denevan WM. 1992. The pristine myth: the landscape of the Americas in 1492. *Ann Assoc Am Geogr* 82: 369–85.

Dether D and Black A. 2006. Learning from escaped prescribed fires – lessons for high reliability. *Fire Management Today* 66: 50–56.

Dickinson MB, Lacki MJ, and Cox DR. 2009. Fire and the endangered Indiana bat. In: Hutchinson TF (Ed). Proceedings of the 3rd Fire in Eastern Oak Forests Conference; 20–22 May 2008; Carbondale, IL. Newtown Square, PA: USDA Forest Service, Northern Research Station. GTR-NRS-P-46.

Engber EA, Varner JM, Arguello LA, et al. 2011. The effects of conifer encroachment and overstory structure on fuels and fire in an oak woodland landscape. *Fire Ecology* 7: 32–50.

Finney MA, McHugh CW, and Grenfell IC. 2005. Stand- and landscape-level effects of prescribed burning on two Arizona wildfires. *Can J For Res* 35: 1714–22.

Gilliam FS and Platt WJ. 1999. Effects of long-term fire exclusion on tree species composition and stand structure in an old-growth *Pinus palustris* (longleaf pine) forest. *Plant Ecol* 140: 15–26.

Goebel T, Waters MR, and O'Rourke DH. 2008. The Late Pleistocene dispersal of modern humans in the Americas. *Science* 319: 1497–1502.

Gude PH, Jones K, Rasker R, et al. 2013. Evidence for the effect of homes on wildfire suppression costs. *Int J Wildland Fire*; doi:10.1071/WF11095.

Guyette RP and Spetich MA. 2003. Fire history of oak–pine forests in the lower Boston Mountains, Arkansas, USA. *Forest Ecol Manag* 180: 463–74.

Haines DA and Sando RW. 1969. Climatic conditions preceding historically great fires in the North Central Region. St Paul, MN: USDA Forest Service, North Central Forest Experiment Station. Research Paper NC-34.

Harrington MG. 2012. Duff mound consumption and cambium injury for centuries-old western larch from prescribed burning in western Montana. *Int J Wildland Fire*; doi:10.1071/WF12038.

Harvey HT, Shellhammer HS, and Stecker HE. 1980. Giant sequoia ecology: fire and reproduction. Washington, DC: US DOI National Park Service. Scientific Monograph Series No 12.

Hiers JK, O'Brien JJ, Will RE, et al. 2007. Forest floor depth mediates understory vigor in xeric *Pinus palustris* ecosystems. *Ecol Appl* 17: 806–14.

Hood SM. 2010. Mitigating old tree mortality in long-unburned, fire-dependent forests: a synthesis. Fort Collins, CO: USDA Forest Service. RMRS-GTR-238.

Howe HF. 1994. Response of early-and late-flowering plants to fire season in experimental prairies. *Ecol Appl* 4: 121–33.

Huntzinger M. 2003. Effects of fire management practices on butterfly diversity in the forested western United States. *Biol Conserv* 113: 1–12.

Hutchinson TF, Long RP, Ford RD, et al. 2008. Fire history and the establishment of oaks and maples in second-growth forests. *Can J For Res* 38: 391–403.

Irwin LL, Fleming TL, and Beebe J. 2004. Are spotted owl populations sustainable in fire-prone forests? *J Sustainable For* 18: 1–28.

Kauffman JB and Martin RE. 1990. Sprouting shrub response to different seasons and fuel consumption levels of prescribed fire in Sierra Nevada mixed conifer ecosystems. *Forest Sci* 36: 748–64.

Kay CE. 2007. Are lightning fires unnatural? A comparison of aboriginal and lightning ignition rates in the United States. In: Masters RE and Galley KEM (Eds). Proceedings of the 23rd Tall Timbers Fire Ecology Conference: Fire in Grassland and Shrubland Ecosystems. Tallahassee, FL: Tall Timbers Research Station.

Keane RE, Ryan KC, Veblen TT, et al. 2002. Cascading effects of fire exclusion in Rocky Mountain ecosystems: a literature review. Fort Collins, CO: USDA Forest Service. RMRS-GTR-91.

Keeley JE. 2002. Native American impacts on fire regimes of the California coastal ranges. *J Biogeogr* 29: 303–20.

Keeley JE, Aplet GH, Christensen NL, et al. 2009. Ecological foundations for fire management in North American forest and shrubland ecosystems. Portland, OR: USDA Forest Service, Pacific Northwest Research Station. GTR-PNW-779.

Kilgore BM. 1973. The ecological role of fire in Sierran conifer forests: its application to National Park Management. *Quaternary Res* 3: 496–513.

Knapp EE, Estes BL, and Skinner CN. 2009. Ecological effects of prescribed fire season: a literature review and synthesis for managers. Albany, CA: USDA Forest Service, Pacific Southwest Research Station. PSW-GTR-224.

Knapp EE and Keeley JE. 2006. Heterogeneity in fire severity with early season and late season prescribed burns in a mixed conifer forest. *Int J Wildland Fire* 15: 37–45.

Knapp EE, Schwilk DW, Kane JM, et al. 2007. Role of burning season on initial understory vegetation response to prescribed fire in a mixed conifer forest. *Can J For Res* 37: 11–22.

Knight TM and Holt RD. 2005. Fire generates spatial gradients in herbivory: an example from a Florida sandhill ecosystem. *Ecology* 86: 587–93.

Krezek-Hanes CC, Ahern F, Cantin A, et al. 2011. Trends in large fires in Canada, 1959–2007. Ottawa, Canada: Canadian Councils of Resource Ministers. Canadian Biodiversity: Ecosystem Status and Trends 2010, Technical Thematic Report No 6. www.biodivcanada.ca/default.asp?lang=En&n=137E1147-0. Leenhouts B. 1998. Assessment of biomass burning in the conter-minous United States. *Conserv Ecol* 2: 1.

Littell JS, McKenzie D, Peterson DL, et al. 2009. Climate and wildfire area burned in western US ecoprovinces, 1916–2003. *Ecol Appl* 19: 1003–21.

Martinez Domínguez R and Rodríguez-Trejo DA. 2008. Forest fires in Mexico and Central América. In: González-Cabán A (Ed). Proceedings of the Second International Symposium on Fire Economics, Planning, and Policy: A Global View. Albany, CA: USDA Forest Service, Pacific Southwest Research Station. PSW-GTR-208.

Maynard EE and Brewer JS. 2012. Restoring perennial warm-season grasses as a means of reversing mesophication of oak woodlands in northern Mississippi. *Restor Ecol* 20: 1–8.

McCaffrey S. 2006. Prescribed fire: what influences public approval? In: Dickinson MB (Ed). Fire in eastern oak forests: delivering science to land managers. Newtown Square, PA: USDA Forest Service, Northern Research Station. Gen Tech Rep NRS-P-1.

McCaffrey S, Toman E, Stidham M, et al. 2013. Social science research related to wildfire management: an overview of recent findings and future research needs. *Int J Wildland Fire* 22: 15–24.

McClain WE and Elzinga SL. 1994. The occurrence of prairie and forest fires in Illinois and other midwestern states, 1679–1853. *Erigenia* 13: 79–90.

McShea WJ, Healy WM, Devers P, et al. 2007. Forestry matters: decline of oak will impact wildlife in hardwood forests. *J Wildlife Manage* 71: 1717–28.

Means DB. 2006. Vertebrate faunal diversity of longleaf pine ecosystems. In: Jose S, Jokela EJ, and Miller DL (Eds). The longleaf pine ecosystem: ecology, silviculture, and restoration. New York, NY: Springer.

Motzkin G, Patterson WA, and Foster DR. 1999. A historical perspective on pitch pine–scrub oak communities in the Connecticut Valley of Massachusetts. *Ecosystems* 2: 255–73.

NIFC (National Interagency Fire Center). 2013a. Prescribed fires. www.nifc.gov/fireInfo/fireInfo_stats_prescribed.html. Viewed 27 Mar 2013.

NIFC (National Interagency Fire Center). 2013b. Wildland fire use fires. www.nifc.gov/fireInfo/fireInfo_stats_fireUse.html. Viewed 27 Mar 2013.

North M, Stine P, Zielinski W, et al. 2010. Harnessing fire for wildlife. *The Wildlife Professional* 4: 30–33.

Noss RF, Franklin JF, Baker WL et al. 2006. Managing fire-prone forests in the western United States. *Front Ecol Environ* 4: 481–87.

Nowacki GJ and Abrams MD. 2008. The demise of fire and "mesophication" of forests in the eastern United States. *BioScience* 58: 123–38.

Nowacki GJ, MacCleery DW, and Lake FK. 2012. Native Americans, ecosystem development, and historical range of variation. In: Wiens JA, Hayward GD, Safford HD, and Giffen CM (Eds). Historical environmental variation in conservation and natural resource management. Chichester, UK: John Wiley & Sons.

Pausas JG and Keeley JE. 2009. A burning story: the role of fire in the history of life. *BioScience* 59: 593–601.

Peters MP, Iverson LR, Matthews SN, et al. 2013. Wildfire hazard mapping: exploring site conditions in eastern US wildland–urban interfaces. *Int J Wildland Fire*; doi:10.1071/ WF12177.

Platt WJ. 1999. Southeastern pine savannas. In: Anderson RC, Fralish JS, and Baskin JM (Eds). Savannas, barrens, and rock outcrop plant communities of North America. Cambridge, UK: Cambridge University Press.

Power MJ, Mayle FE, Bartlein PJ, et al. 2012. Climatic control of the biomass-burning decline in the Americas after AD 1500. *Holocene* 23: 3–13.

Prichard SJ, Peterson DL, and Jacobson K. 2010. Fuel treatments reduce the severity of wildfire effects in dry mixed conifer forest, Washington, USA. *Can J For Res* 40: 1615–26.

Pyne SJ. 1982. Fire in America: a cultural history of wildland and rural fire. Princeton, NJ: Princeton University Press.

Pyne SJ, Andrews PJ, and Laven RD. 1996. Introduction to wildland fire. New York, NY: John Wiley & Sons.

Quinn-Davidson LN and Varner JM. 2012. Impediments to prescribed fire across agency, landscape and manager: an example from northern California. *Int J Wildland Fire* 21: 210–18.

Rodríguez-Trejo DA. 2008. Fire regimes, fire ecology, and fire management in Mexico. *AMBIO* 37: 548–56.

Rodríguez-Trejo DA and Fulé PZ. 2003. Fire ecology of Mexican pines and a fire management proposal. *Int J Wildland Fire* 12: 23–37.

Rodewald AD and Abrams MD. 2002. Floristics and avian community structure: implications for regional changes in eastern forest composition. *Forest Sci* 48: 267–72.

Ryan KC. 2002. Dynamic interactions between forest structure and fire behavior in boreal ecosystems. *Silva Fennica* 36: 13–39.

Ryan KC and Reinhardt ED. 1988. Predicting post-fire mortality of seven western conifers. *Can J For Res* 18: 1291–97.

Ryan KC and Frandsen WH. 1991. Basal injury from smoldering fires in mature *Pinus ponderosa* Laws. *Int J Wildland Fire* 1: 107–18.

Ryan KC, Jones AT, Koerner CL, et al. 2012. Wildland fire in ecosystems – effects of fire on cultural resources and ecosystems. Ft Collins, CO: USDA Forest Service, Rocky Mountain Research Station. RMRS-GTR-91.

Ryan KC and Opperman TS. 2013. LANDFIRE – a national vegetation/fuels data base for use in fuels treatment, restoration, and suppression planning. *Forest Ecol Manag* 294: 208–16.

Schroeder MJ and Buck CC. 1970. Fire weather: a guide to application of meteorological information to forest fire control operations. Washington, DC: USDA Forest Service. Agriculture Handbook 360.

Schwilk DW, Keeley JE, Knapp EE, et al. 2009. The national fire and fire surrogate study: effects of fuel reduction methods on forest vegetation structure and fuels. *Ecol Appl* 19: 285–304.

Show SB and Kotok EI. 1924. The role of fire in the California pine forests. *US Department of Agriculture Bulletin* 1294.

Spetich MA, Perry RW, Harper CA, et al. 2011. Fire in eastern hardwood forests through 14 000 years. In: Greenberg CH, Collins BS, and

Thompson III FR (Eds). Sustaining young forest communities. Dordrecht, the Netherlands: Springer.

Stambaugh MC, Guyette RP, and Marschall JM. 2011. Longleaf pine (*Pinus palustris* Mill) fire scars reveal new details of a frequent fire regime. *J Veg Sci* 22: 1094–04.

Stephens SL. 2005. Forest fire causes and extent on United States Forest Service lands. *Int J Wildland Fire* 14: 213–22.

Stephens SL, Martin RE, and Clinton NE. 2007. Prehistoric fire area and emissions from California's forests, woodlands, shrublands, and grasslands. *Forest Ecol Manag* 251: 205–16.

Stoddard HL. 1931. The bobwhite quail: its habits, preservation and increase. New York, NY: Charles Scribner's Sons.

Streng DR, Glitzenstein JS, and Platt WJ. 1993. Evaluating effects of season of burn in longleaf pine forests: a critical literature review and some results from an ongoing long-term study. In: Hermann SM (Ed). The longleaf pine ecosystem: ecology, restoration and management. Proceedings of the 18th Tall Timbers Fire Ecology Conference. Tallahassee, FL: Tall Timbers Research Station.

Sun C and Tolver B. 2012. Assessing administrative laws for forestry prescribed burning in the southern United States: a management-based regulation approach. *Int Forest Rev* 14: 337–48.

Swetnam TW and Betancourt JL. 1990. Fire–Southern Oscillation relations in the southwestern United States. *Science* 249: 1017–20.

Taylor AH. 2004. Identifying forest reference conditions on early cut-over lands, Lake Tahoe Basin, USA. *Ecol Appl* 14: 1903–20.

Taylor S. 1998. Prescribed fire in Canada…a time of transition. *Wildfire* 7: 34–37.

Vale T (Ed). 2002. Fire, native peoples, and the natural landscape. Washington, DC: Island Press.

Varner JM, Gordon DR, Putz FE, et al. 2005. Restoring fire to long-unburned *Pinus palustris* ecosystems: novel fire effects and consequences for long-unburned ecosystems. *Restor Ecol* 13: 536–44.

Wade D and Brenner J. 1992. Florida's 1990 Prescribed Burning Act: protection for responsible burners. *J Forest* 90: 27–30.

Weber MG and Taylor SW. 1992. The use of prescribed fire in the management of Canada's forested lands. *Forest Chron* 68: 324–34.

Wenger E. 2000. Communities of practice and social learning systems. *Organization* 7: 225–46.

Westerling AL, Hidalgo HG, Cayan DR, et al. 2006. Warming and earlier spring increase western US forest wildfire activity. *Science* 313: 940–43.

Yoder J, Engle D, and Fuhlendorf S. 2004. Liability, incentives, and prescribed fire for ecosystem management. *Front Ecol Environ* 2: 361–66.

PRESCRIBED FIRE POLICY BARRIERS AND OPPORTUNITIES: A DIVERSITY OF CHALLENGES AND STRATEGIES ACROSS THE WEST

Courtney Schultz, Heidi Huber-Stearns, Sarah Mccaffrey, Douglas Quirke, Gwen Ricco and Cassandra Moseley

Summer 2018

About the Authors

Courtney Schultz is the Director of the CSU Public Lands Policy Group and Associate Professor in the Department of Forest and Rangeland Stewardship, Colorado State University.

Heidi Huber-Stearns is Associate Director of the Ecosystem Workforce Program, Institute for a Sustainable Environment, University of Oregon.

Sarah McCaffrey is a Research Social Scientist with the U.S. Forest Service Rocky Mountain Research Station.

Douglas Quirke is a Research Associate at the University of Oregon School of Law's Environmental & Natural Resources Law Center.

Gwen Ricco is a Research Associate with the Public Lands Policy Group, Colorado State University.

Cassandra Moseley is Director of the Ecosystem Workforce Program, Institute for a Sustainable Environment, and Associate Vice President for Research, University of Oregon.

Acknowledgments

This study was made possible with funding from the Joint Fire Science Program (#16-1-02-8). Many thanks to all of our interviewees for participating in this research. Layout and design by Autumn Ellison, Ecosystem Workforce Program, University of Oregon.

All photos public domain from U.S. Forest Service and USDA Flickr sites.

Ecosystem Workforce Program
Institute for a Sustainable Environment
5247 University of Oregon
Eugene, OR 97403-5247-1472
ewp@uoregon.edu
ewp.uoregon.edu

Public Lands Policy Group
Department of Forest and Rangeland Stewardship
Colorado State University
Fort Collins, CO 80523-1472
courtney.schultz@colostate.edu
sites.warnercnr.colostate.edu/courtneyschultz/

University of Oregon

The University of Oregon is an equal-opportunity, affirmative-action institution committed to cultural diversity and compliance with the Americans with Disabilities Act. This publication will be made available in accessible formats upon request. ©2018 University of Oregon.

Executive Summary

We are conducting a project investigating policies that limit managers' ability to conduct prescribed fire on US Forest Service and Bureau of Land Management (BLM) lands in the 11 Western states. The goals for this phase of our work were to understand the extent to which various policies are limiting prescribed fire programs, strategies to maintain and increase prescribed fire activities, and opportunities for improving policies or policy implementation. To understand the diversity of challenges faced and strategies in use across the West, we conducted a legal analysis of the laws and policies that affect prescribed fire programs on Forest Service and BLM lands (available online at http://ewp.uoregon.edu/publications/working) and approximately 60 interviews with land managers, air regulators, state agency partners, and several NGO partners.

Key Findings

Interviewees in most states said air quality was not the primary variable limiting their application of prescribed fire. The exceptions were in Oregon and Washington, where interviewees said that state-level smoke management programs restrict their ability to burn. Respondents in California also said air quality can be a major consideration; however, they emphasized that there are many other factors that are currently limiting their programs that need to be addressed, and they did not suggest a need for changes to state regulatory approaches at this time. No respondent suggested the need for changes to the Clean Air Act. Some additional details include the following:

- In the Intermountain West, people said air quality is a consideration and constraint for all burners, but that available funding and capacity, other land management considerations, and internal agency dynamics were the primary factors limiting their use of prescribed fire.
- Air quality is a more significant consideration in areas near large population centers where there are many sources of pollution in the airshed, areas with poor air quality (e.g., the Wasatch and San Joaquin Valleys), and sensitive populations.
- In Oregon and Washington, there are relatively stricter state standards for regulating air quality, and burners said air quality regulation is one of the major barriers to burning. Both states are in the process of revising their smoke management programs, and revised programs likely will continue to limit smoke from

prescribed fire to standards stricter than those of the federal Clean Air Act.
- Challenges on the horizon include managing air quality for large, multi-day burns and during natural ignitions that are managed for resource benefit.

Lack of capacity and funding, and challenges sharing resources across agencies were the most significant barriers to accomplishing more prescribed fire that we uncovered in our interviews. Additional details include the following:

- Capacity to burn is limited when burn windows coincide with wildfire season.
- Capacity to burn can be limited outside of wildfire season due to loss of seasonal staff, trainings, and other demands on staff time.
- People cite significant problems sharing resources across units and agencies, due to a lack of flexibility associated with budgetary requirements and challenges using agreement mechanisms efficiently and effectively.

Interviewees said there are limited incentives to burn, making leadership and personal investment in burning central to success.

- Committed leaders, according to our interviewees, often find creative strategies to overcome the multiple challenges associated with burning. Successful programs depend on a personal investment from line officers and fire management officers in conducting prescribed fire.
- The current structure of performance measures within the Forest Service could provide stronger incentives to conduct prescribed fire.
- Interviewees across all states also believed risk aversion was an important factor in willingness to burn. At the local level this tended

to reflect concerns about personal liability in case of an escaped fire. At the higher level it tended to reflect political considerations.

Interviewees also cited other challenges, including short burn windows, planning limitations, and other landscape conditions and conservation priorities (e.g., sage grouse conservation, the presence of cheat grass, steep topography) that significantly limit burning in many places. With regard to planning, some suggested streamlined planning requirements and better coordination across agencies, while others noted the need for fire personnel to be present on planning teams to ensure project design supports prescribed fire. In some locations burn windows are short and infrequent; when coupled with capacity limitations, people said it is often difficult to accomplish burning during their available windows.

Opportunities and Successful Strategies

There is no "silver bullet" to increasing prescribed fire, and finding opportunities requires: collaborative, place-specific problem solving; active coordination across air regulators and land managers; and coordination among burners to share resources, communicate effectively with the public, and manage competition in airsheds. Examples of collaborative bodies and strategies that interviewees pointed to include:

- California's Fire MOU Partnership, which is a voluntary group that involves regulators, CALFIRE, federal land managers, and NGO partners. The group is focused on improving understanding of barriers to prescribed fire and opportunities. Working groups within this partnership are examining why burning does not occur on available burn days, and whether this is due to weather, lack of capacity, poor planning, or other variables.
- The Montana-Idaho Airshed Group, which is run by and for major burners (federal and state land managers, and large private landowners) to coordinate burning activities and streamline

communication with the state air quality regulatory bodies. The group tracks all planned burns and communicates on behalf of the burners to regulators.
- Dedicated air quality liaisons and smoke coordinators, who are federal agency employees that work directly and often daily with state air quality regulators. The first such position was in Arizona and was jointly funded by the Arizona DEQ and the Forest Service; the Forest Service has these positions in place in many states. The Department of Interior also has similar positions in some states, but there are opportunities to expand this practice for both agencies.
- People on the ground have some strategies to share resources through agreements and use of the Good Neighbor and "Wyden" Authorities; however, people said that finding easier ways to share resources and charge to common funding codes are high priorities for change.

Conclusion and Recommendations

Our interviews did not yield clear indications that policy change is needed at the federal level at this time, as most interviewees said there were opportunities to increase the use of prescribed fire that would not require changes to federal law. Realizing these opportunities will require creative problem-solving, and a commensurate input of staff time, funding and capacity, and leadership initiative. Two areas where policy change may be warranted are 1) in smoke management programs in Oregon and Washington, where such revisions are underway, and 2) potentially to facilitate easier approaches to interagency resource sharing. In addition, changes to incentive structures within the Forest Service may be warranted, and it is worth exploring possible internal practices that could alleviate current capacity limitations. Some suggestions drawn from our interviews include:

- Ensuring air quality liaisons are in place for all states and exploring whether additional state-level groups, modeled after practices in California and Montana/Idaho, are needed to coordinate among burners and with air quality regulators.
- Improving internal incentives to burn through redesign of some performance measures or the creation of special initiatives with funding that units and collaborative partners could compete for.
- Identifying more efficient and effective avenues for resource sharing. Suggestions include: centralizing contracts and agreements staff, or finding other ways to ensure they are knowledgeable about all options and give consistent advice; creating other agreement mechanisms that are less cumbersome than current options; and finding ways to charge more easily to single budget lines when using resources from multiple agencies. As our work continues, we will explore whether any of these recommendations may require policy changes.
- Ensuring capacity is available through improved strategic planning, use of dedicated prescribed fire crews, greater flexibility to use fire personnel across units, and more effective use of partner capacity.
- Improving measurement of smoke generation and dispersion in order to identify additional opportunities to burn and promote transparency in decision making. Investments could be directed to necessary equipment and meteorologist positions.

This report contains additional, specific details on the strategies in place and suggestions from participants in this phase of our research. Our future work will build on this analysis with case studies in locations that are currently finding ways to build their prescribed fire programs and will include ongoing dialogue with practitioners, partners, agency leadership, and policymakers.

We are conducting a project investigating policies that limit managers' ability to conduct prescribed fire on US Forest Service and Bureau of Land Management lands in the 11 Western states. Our primary objectives are to: 1) Identify current perceived policy barriers to implementing prescribed fire and how these vary across the West, and 2) Characterize actionable opportunities and mechanisms for overcoming barriers. Ultimately, our aim is to identify which policies present the greatest priorities and opportunities for change, and what the mechanisms are for realizing those opportunities. This report details our findings from our initial phases of research on this project, including a legal analysis and approximately 60 interviews with key informants (e.g., land managers, air regulators, and state agency partners).

Recent papers on the challenges in US fire management generally have emphasized the need for policy change to support prescribed fire, and some have suggested there may be a need to reduce regulatory restrictions on smoke to allow for more application of prescribed fire to promote fire-adapted ecosystems and communities (North et al. 2015; Schoennagel et al. 2017). As a result, there is a widely accepted understanding that the current policy environment significantly constrains decision-making around prescribed fire (USDA & USDI 2014).

The term "policy" encompasses a variety of actions taken (or not taken) by government, and changing policy is a complex process. To identify both where policy change may be necessary and also possible, it is critical to distinguish between policy barriers that are: 1) fixed in congressional laws, 2) a result of state or federal agency policy interpretations (e.g., regulations and agency guidance), 3) a result of agency culture or habit, and 4) a result of individual decision-making at the field level, where decisions are influenced by factors such as the social environment in which decision-makers act and their individual degree of risk-aversion (Moseley and Charnley 2014). Each type of policy-related challenge presents different opportunities, risks, and pathways for change. Amending federal environmental law through the US Congress is difficult to achieve; issues for congressional action must be on the political agenda and often require substantial lobbying of members of Congress to champion legislative changes. Regulatory changes can be undertaken by the executive branch, but they also take many years to achieve through rule-making processes under the Administrative Procedures Act and can be amended by subsequent administrations. Substantial changes to agency policy generally are relatively less difficult to achieve, although altering agency norms and behaviors requires sustained changes to communication, leadership, and incentive structures (Fernandez & Rainey 2006). It is also important to note that policy changes may have limited efficacy and unexpected effects. These considerations should inform discussions of policy change as an avenue for increasing the application of prescribed fire on federal lands.

Approach

This project was funded by the Joint Fire Science Program in 2016 with the objectives of identifying the origin and range of interpretation of perceived policy barriers, characterizing the opportunities and mechanisms that are available to overcome barriers at various scales, and educating stakeholders about the most ready opportunities for change. The project involves four primary tasks: 1) a legal analysis of laws that affect the use of prescribed fire (available online at http://ewp.uoregon.edu/publications/working); 2) interviews across the 11 western states to identify the diversity of approaches and challenges associated with accomplishing prescribed fire; 3) a spatial analysis of prescribed fire accomplishments and their correlation with air quality; and 4) case studies of locations that are actively finding ways to accomplish more prescribed fire. This report details our findings from tasks 1 and 2.

We began with an analysis of the major policies that constrain prescribed fire, including a detailed investigation of state-level air quality regulation under the federal Clean Air Act. For the state-level investigation, we initially identified references to prescribed fire, smoke management, and visibility or regional haze in state implementation plans (SIPs), which are required by the Clean Air Act. We reviewed state laws pertaining to prescribed fire and additional state laws, policies, or plans relevant for prescribed fire on federal lands. Subsequent interviews with practitioners generally revealed more specific details regarding the implementation of laws and policies on the ground, and brought to light additional laws and policies having an effect on implementation of prescribed fire. We reviewed these as necessary to complete our legal analysis.

In Fall of 2017, we began interviews across the 11 western states.[10] Our goal was to obtain a broad understanding of policy barriers to prescribed fire across the West, and to identify differences across the states and opportunities for improving practice. We were not conducting comprehensive case studies of every state in this analysis. Our approach was to interview a lead person for the BLM and Forest Service in each state. At the state or regional level for these agencies, we identified people who were fuels program leads or directors/assistant directors of fire and aviation management. We also spoke to air quality or smoke management liaisons within these agencies when our primary point of contact recommended we do so. In states where the Forest Service has no regional office, we spoke to a fire management staff person at the national forest level. We also reached out to state forestry agencies to identify a contact for each state and to state

---

[10] The 11 western states include: Arizona, California, Colorado, Idaho, Montana, Nevada, New Mexico, Oregon, Utah, Washington and Wyoming.

departments of environmental quality to hear the perspective of air quality regulators. In the states where they exist, we also spoke to chairs of prescribed fire councils. In the end, we targeted at a minimum one Forest Service, one BLM, one state forestry, one air quality regulatory, and one prescribed fire council individual for all 11 states. Our total number of interviews was 56, with some state-to-state variation, due either to unwillingness to participate or recommendations for additional, key people to interview. Interview questions focused on: 1) goal-setting processes and progress towards goals for the land management agencies; 2) regulatory processes for regulatory agencies; 3) barriers to improving prescribed fire accomplishments, strategies and suggestions for increasing use of prescribed fire, and 5) the role of partners and communication in supporting the use of prescribed fire.

I. Air quality regulation and prescribed fire

Because the literature had identified air quality regulation as a major barrier to prescribed fire and an arena for potential policy change, we investigated this topic in detail in both our legal analysis and interviews. In this section, we provide an overview of interviewees' perspective on air quality regulation as it interacts with prescribed fire programs (the legal analysis of laws that affect the use of prescribed fire is available online at http://ewp.uoregon.edu/publications/working). Policy in this area is complex, necessitating some background, provided in the next section, on how regulation works under the Clean Air Act in order to interpret our findings. Subsequent sections report on findings from both our legal and interview analyses.

An Overview of Relevant Legal Provisions in the Federal Clean Air Act
Federal Clean Air Act regulation of prescribed fire emissions primarily addresses two categories of potential consequences of such emissions:

- The potential for prescribed fires emissions to violate National Ambient Air Quality Standards (NAAQs); and
- The potential for prescribed fires emissions to negatively affect visibility and regional haze.

States have smoke management programs to maintain compliance with requirements related to both regional haze and NAAQs. The eleven states encompassed by this project generally regulate emissions from prescribed fires for both of these potential effects, with specific details of programs varying from state to state.

Smoke management programs are typically incorporated into state regulatory law, and the elements of a state's smoke management program that are legally binding under the Clean Air Act also are referenced in the State Implementation Plan (SIP).

NAAQs

The federal Clean Air Act requires the federal Environmental Protection Agency (EPA) to establish National Ambient Air Quality Standards (NAAQS), and "states have the primary responsibility for achieving and maintaining" these standards.[11] EPA has established standards for carbon monoxide, lead, nitrogen dioxide, ozone, particle pollution, and sulfur dioxide. States then must outline their strategies for achieving and maintaining the standards for each of these pollutants in their SIPs. Areas within states are designated as in "attainment," "nonattainment," or an "unclassifiable" status based on available information. The major pollutants of concern from fires are particulate matter—both coarse (PM_{10}) and fine ($PM_{2.5}$)—and ozone precursors (NWCG, 2001, as cited in Engel, 2013).

SIPs

State implementation plans (SIPs) are required under the Clean Air Act, are legally binding, and incorporate a range of tools to address air pollution, including statutes, regulations, directives, manuals, and county and municipal ordinances. The Clean Air Act and its implementing regulations (promulgated by the federal EPA) establish minimum standards for SIPs, with differing "requirements and procedures . . . triggered depending on the degree of attainment or nonattainment of the NAAQS."

Visibility and Regional Haze[12]

The Clean Air Act's visibility protection requirements date to 1977 amendments to the Act aimed at remedying existing and preventing future "impairment of visibility" in "Class I Areas," which are primarily designated wilderness areas over 5,000 acres in size and national parks over 6,000 acres in size. There are 108 Class I areas in the eleven-state region encompassed by this project, which amounts to 69% of all Class I areas nationwide. Congress amended the Clean Air Act in 1990 to address impairment of

---

[11] This information is drawn from D. Braddock & Alec C. Zacaroli, Meeting Ambient Air Standards: Development of the State Implementation Plans, in The Clean Air Act Handbook, pp. 49-87 (Julie R. Domike & Alec C. Zacaroli eds. 2016).

[12] For more information, see generally M. Lea Anderson, The Visibility Protection Program, in The Clean Air Act Handbook, pp. 219-248 (Julie R. Domike & Alec C. Zacaroli eds. 2016).

visibility in Class I areas by "regional haze," or "visibility impairment that is produced by a multitude of sources and activities that are located across a broad geographic area"[13] Current regional haze regulations require comprehensive SIP revisions to strengthen existing regional haze SIPs by July 31, 2021. Revised regional haze SIPs must focus on "attain[ing] natural visibility conditions by the year 2064,"[14] and must include "a long-term strategy that addresses regional haze visibility impairment for each mandatory Class I Federal area within the State and for each mandatory Class I Federal area located outside the State that may be affected by emissions from the State."[15]

Exceptional Events Rule: When exceptional events, such as a wildland fire, occur, a state can petition EPA to exclude the monitored emissions of that event from assessments of state compliance with SIPs. In recent years, EPA has signaled increased support for prescribed fires in its revised regulations regarding exceptional events. EPA has stated that it "do[es] not expect the total acreage subject to prescribed fires on wildlands to decrease in the future because prescribed fire is needed for ecosystem health and to reduce the risk of catastrophic wildfires."[16]

Although managers are not allowed to plan a prescribed fire that will violate a state's SIP (e.g., cause an exceedance of NAAQs), the rule allows emissions from qualifying prescribed fires to be excluded from compliance determinations when smoke from prescribed fires leads to unanticipated exceedances, as long as smoke management is employed and the fire is part of a qualifying prescribed fire program included in a land or resource management plan.[17]

Prescribed fire air quality permitting processes

Every state is unique in its regulatory structure and interagency partnerships for overseeing air quality impacts from prescribed burning (see Table 1, page 14 for an overview of legal requirements by state). Most states

---

[13] 5 81 Fed. Reg. 26,942, 26,946 (2016).
[14] 6 40 CFR § 51.308(d)(1)(i)(B).
[15] 7 40 CFR § 51.308(d)(3).
[16] 8 81 Fed. Reg. 26,959 (2016).
[17] 9 40 CFR §§ 50.1(j),(k) & (m) through (r), 50.14, and 51.930.

have a Department of Environmental Quality (DEQ) or equivalent office that handles air quality permitting for prescribed burning. Exceptions include: California, where the California Air Resources Board oversees 30+ air pollution districts or control boards that handle permitting for specific areas; Nevada, where two county offices handle permits for their county, while the NV Department of Environmental Protection handles permits for the rest of the state; Oregon, where the Oregon Department of Forestry handles permitting as a conduit between the Oregon DEQ and burners; and Washington, where the Department of Natural Resources handles permitting for federal public lands.

All states have unique permitting processes that depend on their smoke management plans, regulatory structure, and local considerations. Some states, like New Mexico and Wyoming, have a permit-by-rule system, whereby burners must register burns and notify DEQs about burning activities, but do not receive a permit. In Colorado and Washington, air quality agencies write permits for each burn plan, usually with daily acreage limits that vary depending on ventilation conditions. In other states, such as Montana and Idaho, the DEQ writes a single permit for the entire year for each "major burner," a category that includes each land management agency.

During much of the burn season, daily coordination calls are held between DEQ and with major burners to minimize conflicts and potential smoke impacts. In Arizona, burners register their burns and smoke management prescriptions with the DEQ annually and then must seek a daily permit, based on daily conditions and considerations. Permitting in California proceeds similarly, with annual registration of planned burns in the Prescribed Fire Information Reporting System and then a daily coordination call to communicate whether burning is allowable on a particular day and for coordinating and approving planned burns within 24 hours of ignition. States generally require 24-hour-prior notification of plans to burn and postburn reporting.

Air Quality as a Barrier to Prescribed Fire

Although air quality is a consideration and constraint for all burners, many interviewees, particularly in the Intermountain West, said air quality is not the primary barrier they face to increasing their prescribed fire accomplishments. As we discuss in more detail in Section II of this paper, in most of the states in the Intermountain West, people said that available capacity (resources and personnel), other land management considerations, or internal agency dynamics were the primary factors limiting their use of prescribed fire. People acknowledged that while there are times when air quality is a limiting factor, there are often many other days they can burn. Some staff indicated that if they were burning as much as they should be to mimic natural ecological processes, then air quality would become a major consideration; however, people said their programs were nowhere near this ambitious, because of other reasons like risk tolerance, funding, capacity, and competing priorities. When we asked why air quality gets highlighted as a barrier, interviewees indicated that an air quality permit is an easy variable for managers to focus on, because it is a structured process and often the last piece of the puzzle to put into place in planning a burn. To illustrate, we include here a sample of comments from different land managers:

- "We have worked really hard to communicate and build relationships with our air quality folks in Arizona and New Mexico. I think there are a lot of other things that come into play before air quality does, to keep us from implementing prescribed burns."
- "There's a misconception out there a lot of times that I hear, that the air quality regulator is the barrier that's restricting us from being able to accomplish our burns that we are required to do. I find that is an easy go-to, but the data that we have does not reflect that."
- "The law doesn't necessarily impede prescribed burning so much as some of the more practical realities on the ground. You don't have enough money, you don't have enough people, there's too much fire danger."
- "I think the biggest thing is burn window availability. The smoke side of it . . . it does have an effect, but I think it's minor."
- "Air quality is something we have to consider, but it's also just a matter of, 'Do we have the people to burn where we want to burn? Do we have the burn windows? Is there political tolerance?' I've heard from a number of people that they feel like air quality gets almost scapegoated as an easy excuse sometimes. I'll say . . . it does get scapegoated, because it has a structure that you have to follow."
- "Air quality plays a role in all these things, but in my experience people like to complain about it. But, I haven't seen it deemed a major barrier. Once people have all their ducks in a row and are ready to go, air quality is generally not the issue."
- "I think a lot of people kind of hang their hat on [air quality permitting] being our major implementation barrier, but when you start to look at the numbers, I don't think it's the major one. It's definitely a component that restricts...kind of narrows our windows when we can use prescribed fire . . . extra hoops that we have to jump through. And it's not universal [i.e., it's different from state to state]."

Air quality is a more significant consideration in areas with large populations centers, poor air quality, and sensitive populations. Being close to Class I airsheds or population centers, where there are many sources of emissions that compromise air quality, presents both land management agencies with more air-quality related considerations. For example, one person explained that air quality was a challenge on Colorado's Front Range, given population centers and the presence of a Class I airshed (Rocky Mountain National Park). As another person noted, "When you go to a national park...the one time in your life you might visit an individual park, you can have a very poor experience because of fire It's really hard to convince somebody this is a wonderful, natural experience." Inversions in places like Missoula or the Wasatch Valley of Utah were cited as a limiting factor, as was air quality in highly populated and polluted areas of California, such as the San Joaquin Valley. In Oregon and Washington, in addition to relatively higher levels of regulation, which we discuss below, some said towns with high levels of tourism and smokesensitive populations can be less tolerant, leading land managers and air quality regulators to be more careful about smoke intrusions[18] than the NAAQS would require. Another person pointed to communities throughout Arizona with people who have moved their specifically because they are sensitive to smoke and air pollution. However, outside of the West Coast states, people did not indicate these considerations were primary variables limiting their burning programs.

In Oregon and Washington, there are relatively stricter state standards for regulating air quality; in these states, burners said air quality regulation is one of the major barriers to burning. Both states limit smoke intrusions into communities, even in cases where these would not cause an exceedance of a NAAQS. For example, a prescribed fire might result in a temporarily unhealthy level of smoke that the state regulator deems intolerable even when it might not trigger an exceedance if the NAAQS is based on a 24-hour standard. One person explained, "Washington really has been strict. They don't want any intrusion of any

---

[18] Smoke intrusion: "smoke from prescribed fire entering a designated area at unacceptable levels" (NWCG, 2012).

smoke into any communities at any time." Prescribed burns are generally prohibited on weekends (Friday-Sunday) between June 15 and October 1 in Washington (though there are provisions allowing for exceptions to this prohibition). Another person explained that smoke management plans and permitting in Washington also create barriers to burning, saying "when it comes to air quality regulation, the biggest barrier is the way the smoke management plans and the permitting [are] implemented [which] is really [about] protection against short term intrusions of smoke or nuisance smoke." They went on to explain that even if federal standards are not violated, it can lead to complaints from the public, discussions of fines from the state, and increased local regulation. An interviewee said in Oregon they would like to see ongoing consideration of sub-24-hour intrusions but less formal regulation to a standard that exceeds that of EPA. A number of people said that inversions and intrusions tend to happen at night, even during times of good daytime dispersion, limiting the ability to burn. People indicated that the tolerance of individual regulators in Oregon for writing intrusion reports and dealing with public backlash leads to variability in what is allowed across the state. When discussing tradeoffs between human health and then need for fire, it was in these two states that burners consistently said there was a need to improve smoke management plans, noting that some of these changes were in the works.

Both Oregon and Washington are revising their smoke management plans, which will require demonstration to EPA that changes to regulation will not lead to a greater chance of an exceedance of a NAAQS.

In California, multiple sources of pollution and high population levels can lead to air quality conditions that restrict burning. One person explained, when discussing communication with Air Pollution Control Districts, "Some of these air districts have taken...more restrictive policies than the law requires. Some of those air districts might loosen up those policies. But, in California, if you are burning in an area where your smoke is going to wind up in the Central Valley, it's always going to be difficult, because you're dealing with so much competition for your air. The farming industry, manufacturing, cars, diesel trucks . . . everybody wants to pollute We're the easiest tap to turn off." Another person said, "We do face challenges on air quality, but we've sort of submitted to those challenges, if you will We're competing with folks who are burning wood smoke in their old wood fireplaces. A couple million people doing that every day." Despite this, people did not highlight a need for regulatory change, but rather the need for more communication and creativity to help identify opportunities within the current legal framework. One person said, "The air regulations are going to be an impediment . . . , but I feel like there's a little bit of change happening Particularly after a year of really large, catastrophic wildfires, and the [fact that the] science shows that prescribed fire under almost all conditions produce[s] significantly less smoke per acreageI feel like the air regulators are really working with us, but we are going to continue to comply with the statutes, as they exist." One person suggested changes to air quality regulations may be needed in the future, but everyone said that, before focusing on changes to regulations designed to protect human health, there were other priorities to address to increase use of prescribed fire, including better monitoring and planning to find ways to burn without triggering the NAAQS, addressing capacity issues, and planning more strategically to capitalize on burn days when they are available (see the Section II for more information on these topics).

Additional Themes Around Air Quality

Land managers and air quality regulators both discussed the importance of air quality regulation for protecting human health. Many regulators emphasized their professional duty to protect sensitive populations from air quality risks; in states with smoke management liaisons, who work for the land management agencies and interface with the DEQs, those individuals often also expressed significant concerns about the public health impacts of prescribed fire. People said a fundamental challenge is determining what is an acceptable level of risk to public health from prescribed fire. One regulator said, "One hour of the wrong smoke level can trigger an asthma attack [and] put someone in the hospital . . . . That's my main concern . . . are those vulnerable populations who can't really afford to protect themselves."

Land managers also often acknowledged that air quality can be a life or death matter for individuals, and that the NAAQS may not be protective enough for sensitive populations. Another person from a land management agency explained, "One of the first things that I always talk to, when I talk to [staff frustrated with air quality regulation]—the first thing I explain to them is if we waited for an exceedance on burns, there would be people that would probably die."

Interviewees said allowing prescribed fire requires trust that land managers are doing their best to limit smoke impacts and that prescribed fire will prevent wild-fires in the future. One person noted, "I think the law has tried to facilitate prescribed burning, but not really give a blank check." Several people emphasized the need for air quality regulation and said that land managers, with their professional training and incentives oriented towards land management objectives, could not be relied upon to manage for smoke without input from air quality regulators, who are focused on and trained to address human health considerations.

Larger-scale, landscape burning is particularly challenging to achieve and to permit from a regulatory perspective. Some people explained that it is difficult to find multiple days in a row with the right weather conditions, adequate capacity, and air quality/dispersion conditions to facilitate large burns. From an air quality permitting perspective, it can be both uncertain and risky to permit large burns that may go on for weeks. In some places near towns, where smoke settles into populated areas at night, some people suggested landscape burning is difficult to justify and achieve, given the risks.

In California, in particular, this issue may require attention. One person explained, "So we've all been saying, in all of our venues where we come together with air regulators, we need longer windows, and we need more opportunity to burn on marginal days. We've got to expand the permission space.

And we don't mean that to hurt anyone, to cause them to go to a hospital, or because we don't care, or anything like that; it's just that there's an emissions trade off every time we don't burn that we need to call out as very much a real thing. It's not speculative anymore. California is so flammable these days that we're trying to push this conversation . . . are the [burn] windows long enough? No, they're not."

The argument that communities have to face smoke now (i.e., through prescribed fire) or later (i.e., through wildland fire) was not convincing to many on the ground.

Regulators emphasized that this argument hinges on the assumption, which may not always be true, that prescribed fire now will limit wildland fire later. Others noted that people in general prefer to put off risks into the future, particularly when those risks may never come to pass. Some noted that a key difference between wildfire and prescribed fire is that prescribed fire is intentionally lit, and, therefore, the government has a responsibility to minimize harm in a way that fundamentally differs from a wildland fire event.

Ultimately such dichotomies are too simplistic to accommodate the deliberative dialogue about prescribed fire so many emphasized as being critical among the public, land managers, and air quality regulators. However, several people emphasized that smoke from prescribed fire, which can be done under controlled conditions with good ventilation, is far preferable to smoke from wildland fire. The challenge is ensuring that all involved parties believe the risks of prescribed fire, which may need to be done every few years, are worth taking in order to lower the risks associated with future fire.

Some have suggested that prescribed fire be treated as an exceptional event like wildland fire and not be regulated; this is not a feasible recommendation, according to our interviewees. Although a few interviewees indicated that the new exceptional events rule creates more space to petition for a prescribed fire that causes exceedances of NAAQS to be considered an exceptional event, interviewees also noted that the significance of the rule change was limited because it does not allow prescribed fire to be exempted from regulation. It is not permissible under the Clean Air Act for federal land managers to intentionally plan and cause for exceedances.

As one person said, "The problem with the exceptional events rule is you've gotta have an exceptional event. You can't plan to have an exceptional event." Changing this would require an amendment to the Clean Air Act.

People who spoke to this question said this is not desirable, offering multiple reasons: 1) air quality regulation to stay below NAAQS exceedances is not the biggest barrier to prescribed fire, 2) it would introduce considerable risk to a major environmental law to open it up to amendment, 3) it is unreasonable to think that land managers acting alone will steward air resources with adequate care, finding the ideal balance of burning to reduce risk while protecting human health, and 4) it is politically not viable to look for legislation where a federal land management agency wants an exemption from environmental law in a way that would compromise human health. One person said, "I think politically that would be suicide...public opinion would hang us. [They'd think] the government is trying to kill us."

There is potential for conflict around how smoke from managed natural ignitions is handled; some of these issues may require attention going forward. One regulator noted, "So if they get a natural start...they are going to be putting fire on the ground to keep that fire going as long as they can to avoid having to comply with our requirements because we did not see this coming. They're using that as a way to avoid our requirements for smoke management."

This person explained that avoiding direct communication will only force regulators to act to protect public health. As a separate issue, some discussed that managers can count wildland fire acres burned as accomplishments towards fuels targets; however, in one state, we were told that these acres can only be counted towards targets on days when air quality regulators also would permit burning.

On this topic, one person said, "[Regulators] realize they can't force a suppression. Then you get this policy jockeying around . . . you know, [air quality is] not favorable today, so it's not considered a resource benefit . . . but tomorrow [it] might be. It doesn't change on the ground generally, so it is bizarre." Some of these details may require additional attention to find positive paths forward.

Table 1. State-by-state overview of air quality regulatory process and interagency relationships to support burning

	Regulatory overview: Responsible agencies and applicable law	Prescribed fire planning and approval[19]
Arizona	Arizona Department of Environmental Quality (ADEQ) Arizona Administrative Code	Land managers must make best efforts to register all planned burn projects before December 31 each year, but no later than January 31 ADEQ required to hold meeting after January 31 and before April 1 between ADEQ and land managers to evaluate program and cooperatively establish "annual emission goal" ("planned quantifiable value of emissions reduction from prescribed fires and fuels management activities") Land managers must submit burn plans to ADEQ at least 14 days before burn date Daily burn request must be submitted to ADEQ by 2 P.M. on business day preceding burn ADEQ approval of request required before ignition, with constructive approval where explicit approval is not received from ADEQ by 10 P.M. on the day request was submitted (burner must make effort to confirm that request was received by ADEQ)
California	California Air Resources Board and California's 35 air districts Smoke Management Guidelines for Agricultural and Prescribed Burning (codified in California Code of Regulations)	Smoke management programs for air districts with "prescribed burning in wildlands or urban interfaces" must include annual or seasonal registration of all planned burn projects; burns are registered online in Prescribed Fire Information Reporting System (PIFRS) Each of California's 35 air districts must have a smoke management program that includes a daily burn authorization system Air districts' burn authorization systems issue "48-hour forecasts, 72-hour outlooks, and 96-hour trends" for burns Air district burn authorization systems must include procedures "for authorizing . . . prescribed burns 24 hours prior to ignition" By 3 PM each day, California Air Resources Board must normally announce whether following day is a "permissive burn day" or a "no-burn day" for each of California's 15 air basins

---

[19] Quoted material in this column is drawn from the applicable law indicated for the state in column 1.

Table 1. (Continued)

	Regulatory overview: Responsible agencies and applicable law	Prescribed fire planning and approval
Colorado	Colorado Air Quality Control Commission Colorado Department of Public Health and Environment or an authorized local agency Colorado Code of Regulations Colorado Smoke Management Program Manual	Significant users of prescribed fire must submit planning documents to Colorado Air Quality Control Commission for each area in which the user intends to use prescribed fire addressing the use and role of prescribed fire and resulting air quality impacts Air Pollution Control Division of Colorado's Department of Public Health and Environment must review planning documents and present comments and recommendations to the Commission Commission must hold a public hearing and complete review within 45 calendar days of receipt unless significant user of prescribed fire agrees to longer review period APCD may take up to 30 days to review permit application "Notification of Ignition" must be submitted 2 to 48 hours before ignition "Daily Actual Fire Activity" report due by 10:00 AM on business day following each proposed ignition day
Idaho/Montana	Montana/Idaho Airshed Group with Missoula-based "Smoke Management Unit" that coordinates/administers Idaho and Montana DEQs and local regulatory authorities also have roles Montana/Idaho Airshed Group MOU committing to agreed-upon smoke management program and operating guide	Preseason burn lists entered into Airshed Management System between Dec 1 and Feb 28 for Spring Season burns (March 1 to May 31) and between June 1 and Aug 31 for Fall Season burns (Sep 1 to Nov 30) "Burns that will require more than one consecutive day of ignition to complete require additional coordination" "Special notification and direct approval from both DEQs" required for "Extended-duration Landscape-scale Prescribed Burns" ("ignited and managed over weeks of time to mimic the natural progression of fire on the landscape within parameters identified in the burn plan" and "monitored, additionally ignited, or partially extinguished until season-ending precipitation puts them out completely") Smoke dispersion forecasts posted to Airshed Group web page by approximately 10:00 am Mon through Fri Burns proposed via Airshed Management System by noon day before proposed burn (noon Fri for Sat/Sun/Mon burns) after reviewing dispersion forecast

	Regulatory overview: Responsible agencies and applicable law	Prescribed fire planning and approval
	Idaho and Montana DEQ regulations	Idaho and Montana DEQs and local air agencies "may review the forecast and burn proposals by 2:30 pm . . . and relay any issues or concerns"
		Restrictions/burn recommendation posted by 4 pm
		"Local regulatory authorities . . . may impose additional burn restrictions after the . . . burn recommendations have been posted"
Nevada	Nevada Division of Environmental Protection (NDEP) for all of state except Clark County and Washoe County, which administer program in their jurisdictions Nevada Revised Statutes Nevada Smoke Management Program	Permit application must be submitted at least 30 days prior to planned ignition date for fires emitting more than 10 tons of PM10 Permit application must be submitted at least two weeks prior to planned ignition date for projects emitting between 1.0 and 10 tons of PM10 Land managers must notify the Division as soon as practicable, but no later than 2 pm of the business day preceding the burn Division must issue final decision on the burn (approval, approval with conditions, or disapproval) by 5 pm on the business day prior to ignition or burn is deemed approved Notification to relevant regulatory authorities is required prior to ignition for projects that emit more than 10 tons of PM10 and are within 15 miles of the state border, BIA trust lands managed under the jurisdiction of a tribal air quality agency, or the borders of Washoe or Clark counties
New Mexico	New Mexico Environment Department New Mexico Administrative Code	Different requirements for burn projects with < 1 ton PM-10 emissions per day (SMP-I) and burn projects with one or more ton PM-10 emissions per day (SMP-II) SMP-I: • Notification of populations w/i one mile between 2 and 30 days prior to ignition • Registration by 10 am one business day prior to planned ignition SMP-II: • Registration by two weeks prior to planned ignition • Public notification between 2 and 30 days prior to ignition for burns within 15 miles of a population or w/ wind blowing toward a population

Table 1. (Continued)

	Regulatory overview: Responsible agencies and applicable law	Prescribed fire planning and approval
		• Notification to Dept. between 7 days prior to ignition and 10 am one business day prior to planned ignition Notification of local fire authority prior to ignition required for both
Oregon	Oregon Department of Forestry Oregon Department of Environmental Quality Oregon Administrative Rules Operational Guidance for the Oregon Smoke Management Program	Land managers must register burns with the State Forester at least seven days before the first day of ignition (requirement may be waived if federal policies met) Land managers may request special forecast and instructions at least two days in advance for multi-day burns and burns with > 2,000 tons of fuel loading Smoke Management Forecast Unit issues daily forecasts and instructions no later than 3:15 p.m. during periods of substantial prescribed burning (forecasts and instructions are for the day following issuance) Land managers must provide location, method of burning, and fuel loading tonnages to Smoke Management forecast unit by the day of the burn Land managers must obtain current smoke management forecast and instructions prior to ignition and must conduct burn in compliance with instructions Land managers must follow land management agency policies that provide for affirmative "go-no go decision" before ignition as documented and approved by line officer
Utah	Utah Department of Environmental Quality Division of Air Quality Utah Administrative Code	Director of Utah Department of Environmental Quality's Division of Air Quality must provide opportunity for an annual meeting with land managers to evaluate and adopt annual emission goal, which must be developed in cooperation with states, federal land management agencies and private entities to control prescribed fire emissions increases to the maximum feasible extent; goal is established prior to the beginning of fire season, either at the beginning of the calendar year or before the year begins Land managers must provide director with "long-term projections of future prescribed fire activity" and "list of areas treated using non-burning alternatives to fire during the previous calendar year" by March 15; land managers planning prescribed fire that will burn more than 50 acres annually must also submit a "burn schedule" at this time

	Regulatory overview: Responsible agencies and applicable law	Prescribed fire planning and approval
		Land managers must submit pre-burn information to director for approval at least 2 weeks before beginning of the "burn window" Land managers must submit burn requests for large prescribed fires to the director by 10 AM at least two business days before planned ignition time
Washington	Washington Department of Natural Resources Washington Department of Ecology Smoke Management Plan codified in Washington Administrative Code	Multiple day burns require landowner to give burn plan information to DNR for review three months before the burn, with DNR notification of any additional requirements two months before the burn If DNR determines that the burn has potential to affect communities, landowner must notify public of the burn at least one week before they plan to burn Approval process for "large prescribed fires" (those with potential to create significant smoke impacts beyond the immediate fire area) Land managers responsible for gathering and entering pre-burn site data into smoke management reporting system Land managers screen, pre-authorize/pre-approve and prioritize burns daily, and submit prioritized pre-approvals to Smoke Management Section via Forest Service/DNR data exchange process Smoke Management Section approves or disapproves each burn Land managers give final approval to burns (taking into consideration a list of factors)
Wyoming	Wyoming Department of Environmental Quality's Air Quality Division Wyoming Smoke Management Standards and Regulations (codified as Chapter 10 of Wyoming Administrative Rules)	Burners/land managers "whose total planned burn projects in a year are projected to generate greater than 100 tons of PM10 emissions" must submit written reports to Administrator of Wyoming Department of Environmental Quality's Air Quality Division "by January 31 every third year"; reports must "include documentation of ... long-term burn estimates for the next three years, including the location, burn area or pile volume, vegetation type, and type of burn for each planned burn project Burns projected to generate ≥ 2 tons/PM10 per day (classified as "SMP-II") must be registered with Air Quality Division at least 2 weeks prior to ignition

Table 1. (Continued)

Regulatory overview: Responsible agencies and applicable law	Prescribed fire planning and approval
	Public notification required at least 2 days prior to ignition notification to Air Quality Division 1 hour prior to ignition for SMP-I burns and by 10 A.M. on business day prior to ignition for SMP-II burns Notification to relevant "jurisdictional fire authorities" prior to ignition

II. The Most Common Barriers to Prescribed Fire: Incentives, Capacity, and Conditions on the Ground

Interviewees across all states said internal agency variables, such as funding, incentives, and capacity, tended to pose larger barriers than air quality concerns. In this section we discuss the factors outside of air quality that prevent land management agencies from reaching their prescribed fire goals.

Capacity Challenges: Personnel and Funding

People often said inadequate funding and capacity to accomplish more prescribed fire were the key barriers to accomplishing more burning. Many people made statements like, "My biggest barrier right now is funding," or "We just didn't have the resources." In particular, people said they lacked the funding needed to hire qualified staff to prepare for, plan, and conduct the burns. As one person summarized, "It takes a lot of work to go from planning and doing the NEPA to implementation. We're pretty limited as far as the number of personnel we have." One person added that they often focus their limited budgets on mechanical thinning and explained, "Mechanical work is expensive. So, if we're spending our money on mechanical, then we don't have money to do the final treatments of doing burning on landscape. And, so, the constant push for new mechanical acres then causes a backlog in prescribed fires." Forest Service interviewees across regions felt that the size of the fire suppression budget as proportion of overall agency budgets restricts the amount of burning that can occur. BLM interviewees stated that to plan at landscape scales, units would require more stable funding. Burners with the BLM in states without sage grouse populations said their ability to burn had been limited particularly in recent years, because the agency at the national level had reallocated budgets to states with sage grouse. People said decreased state funding for DEQs also limits regulators' their ability to observe on burns or interact with land managers, which, as we note below, is important to finding opportunities to increasing burning.

Lack of sufficient qualified staff to conduct burns was a key capacity limitation.
- Capacity to burn is limited when burn windows coincide with wildfire season. Across agencies and states, individuals consistently said when wildfires are burning, their qualified personnel leave local units to work on wildland fires. Sometimes when the nation is at a preparedness level four or five, people said it is too difficult to

request fire personnel to work on prescribed fire. One person explained, "One of our big strategic issues is of course when we need to burn in the summer, everybody's fighting wildfire... We just don't have people around to burn. I got certified as a helicopter manager because when I needed to burn in the summer, everybody was gone, doing suppression, and if...I could manage the helicopter, we could burn." One interviewee said that they sometimes can get burning done with severity resources (i.e., people relocated to an area in preparation for wildland firefighting), but this is challenging because those personnel might be called onto a fire at any moment. People said the fact that wildfire seasons are getting longer has exacerbated this problem.

- Capacity to burn can be limited outside of wildfire season due to loss of seasonal staff, trainings, and other demands on staff time. Often land managers want to burn in the off-season when they no longer have seasonal employees to implement burns. One person explained, "Just as burn season is gearing up, we lose most of our workforce. If that didn't happen, I think we would be in a very different position to do landscape-scale burning." In another state someone said, "One of the biggest restrictions is just funding in general. And then, because a lot of our firefighters are more of our operational staff, or more on a seasonal basis, a lot of times they'd be committed to other projects . . . committed to some seeding [elsewhere] or committed to doing some fencing. And so, then . . . when the burn window does open up, we don't have the capacity to complete the objectives, because we don't have the bodies." A few interviewees said that trainings and leave during the holiday season also limit the availability of personnel during key burn windows, especially in the Southwest.

- Some pointed to the challenge of hiring and training qualified burners and "fire adapted" line officers. People said the professionalization of fire personnel has limited the number of people who are available to staff burns. Some interviewees felt that there is a significant challenge in hiring personnel and having the right person in the right position in order to implement prescribed burn programs. One BLM interviewee said, "It's very challenging to hire fuel specialist(s) at the GS9 level[The] field offices are competing with the Forest Service and with [the state forestry agency] for the same personnel. [Those] agencies are often hiring at higher grades My first challenge is personnel-having the right person and the right position, in order to implement these prescribed fires." A Forest Service interviewee pointed out that there was a need, not just for people qualified to conduct a burn, but for line officers who understood fire, explaining a need for "actively finding and developing fire adapted line officers. And, that doesn't mean that they had a lot of fire experience, but that they have a lot of fire knowledge and have people that they can work with and trust to build that knowledge and continue to be able to do fire." One state forestry interviewee shared that their agency "does not hire foresters nor do we have a training program for foresters to be equipped to conduct prescribed burning on the landscape." In multiple states, we heard that if the state land board or forestry division does not support prescribed fire, this can limit federal burners' ability to burn, because it becomes more difficult to share resources, coordinate communication, or work across jurisdictions.

Capacity Challenges: Resource Sharing and Logistics

People cited problems sharing resources across units due to lack of flexibility associated with budget requirements and limitations on travel. One person told us that in the past year they had observed that seasonal fire personnel on a particular forest were inactive, but were not being shared with other forests. When we asked why they said, "I don't know if it's a cultural thing, I don't know if there is actual legal barriers, or the budgets, or whatever it is. There has to be some reason. I know when I talked with people in the past, it's ultimately they have people that they need to give paychecks to. There is a fear of if they start moving around like that, they will lose their budgets. They'll lose their people. They'll lose the ability to pay salaries." Another person indicated that when burn windows fall at the end of the fiscal year, this can be challenging because of the availability of accessing funding as the fiscal year ends and begins. Limitations on travel also have affected the ability to find capacity, according to several interviewees.

People said entering into agreements to share resources is a persistent challenge, and that there is a need for more knowledge of funding mechanisms, streamlined legal advice about their use, and more staff capacity to administer agreements. Many noted that they have to combine the resources of multiple units or agencies to conduct burns, and most people highlighted the challenges associated with sharing resources. One person explained this in detail saying, "We often reach out to our neighboring agencies for assistance with resources and staffing. And that's all facilitated through agreements that we have, both with our state and other federal partners, and that process of getting those agreements in place is often cumbersome. Some of the [authorities] I think are not clearly understood … There [are] differences of opinion between individual grants and agreement specialists, or different lawyers, but there isn't even agreement from one region to another [about] how things are being interpreted. Or when the Washington office, when their staff puts together agreements, they may do something that our staff here says we can't do. And so, there's a lot of inconsistency or different interpretations of how law is applied to these agreements or the authorities that facilitate these agreements … I see that as another big regulatory barrier that exists for us to be able to move forward and further utilize prescribed fire as a management tool." To share resources, often people have in place many agreements with partner agencies. For instance, a Forest Service Regional Office might have two agreements with a corresponding National Park Service unit— often one for outgoing and one for incoming funds, each only lasting five years and requiring tracking and reporting.

A consistent theme was the need to find ways to facilitate more nimble resource sharing, particularly among federal agencies. One person commented, "We try to partner, whether it's with the Forest Service, Fish and Wildlife, [National Park Service], [or Bureau of Indian Affairs], and we're trying to increase the size of the burn—do cross boundary type work. There's no good way to move money between the federal agencies for this. It would really help, because a lot of times, at least in [this state], the Forest Service is the ones with hot shot crews and the helicopters. We want to do larger landscape type burns, and want to use their helicopter. They're more than happy to work with us on that, but it is a nightmare to try and pay for that helicopter . . . . We have to be able to move money between the agencies, just like in a wildfire, we all charge to the same code. Why doesn't that happen on a prescribed fire? It's a huge hindrance." Staff frequently said things such as, "There has to be a way that we can exchange money between the agencies to get these larger landscape burns done." Another person said finding a way to use something akin to the funding system in place during wildland fire to order and pay for resources from other federal agencies would be "the single biggest breakthrough" she could imagine that would help the federal agencies get more fire on the ground.

Agency Leadership and Incentives

Interviewees said there are limited incentives to burn, making leadership a critical component of successful programs. Federal agency employees in particular said burn programs rely on the commitment of agency leadership and fire management officers (FMOs). As one person explained, "I really don't think there's a lot of incentive within the organization to do prescribed fire. I think the incentive comes from the agency administrator and burn boss passion for doing what's right on the landscape." Interviewees said when line officers and FMOs exhibit initiative and passion for burning, they often find creative ways to maintain and build their programs. One interviewee also reflected that local line officers or fire personnel can create a culture supportive of prescribed burning on their units. As they described their forest staff, this person explained, "They have a great deal of enthusiasm and understanding about why this work is important, and how they can use prescribed fires in the future to make good use of wildfire opportunities. There's a lot of focus on being strategic with the use of prescribed fire, and then also with the use of wildfire, and there's a huge amount of support from the regional leadership all the way down to get there. I think the leadership plays a huge role in that."

People said agency history, culture, and professional expertise can all influence prescribed fire programs. Besides the need for clear and active support of senior leadership for prescribed fire, people noted that having a culture or history of focusing on suppression can affect an agency's activities. One BLM employee explained that the agency has more experience, history, and personnel trained for suppression, creating some bias towards suppression and less expertise in prescribed fire. A Forest Service employee noted, "A key part is it's a cultural thing. I think [from] a lot of places [where] we get our fire folks, they come from the suppression background, so suppression is what they know . . . . They may not be completely comfortable with prescribed fire."

The structure of agency performance measures creates weak incentives to use prescribed fire. Setting fuels targets (i.e., acres on which fuel loads have been reduced by a certain amount) can create incentives for land management agencies to increase prescribed fire use. However, mechanical treatments (i.e., removal of fuel through mechanical thinning) can be a more predictable way to meet fuels targets with less associated risk, both to the public and to agency staff who need to implement projects and meet targets. As one person said, "[M]echanical treatments typically have wide open windows ... [they] can happen 10 months, 11 months of the year, versus prescribed fire on a specific piece of ground. You may only have a few days here and there . . . to put that [prescribed fire] project on the ground." Others noted that the timber target (i.e., volume of board feet sold) is more challenging to meet than fuels targets, and that timber targets have gone up for the Forest Service in recent years. Mechanical thinning can help managers achieve both targets, while prescribed fire only contributes to the fuels target. Several Forest Service personnel noted that it is not difficult to meet fuels targets, particularly because they can count wildland fire acres that burn for resource benefit towards their targets, leading to relatively more emphasis on meeting timber targets (however, this has changed as of FY 18; regions and forests now only count prescribed fire and non-fire treatments towards their targets, although acres treated through natural ignitions that burn for resource benefit are still counted at the national level). A BLM employee raised another dynamic around increasing prescribed burning, explaining, "I don't want them pushing getting prescribed fire work done to meet our target, because once we start doing that, then we can end up putting fire on the ground when we shouldn't be."

Interviewees across all states also believed risk aversion was an important factor in willingness to burn. At the local level this tended to reflect concerns about personal liability in case of an escaped fire. At the higher level it tended to reflect political considerations. One interviewee explained: "It gets to that risk aversion component with our line officers or even our burn bosses. And I would say with the burn bosses . . . it probably gets more back into those liability questions, tort claims, and the potential consequences if there's an escape that's created some risk aversion with our implementers for sure. I think at the agency administrator level, it's probably more the social/political components that create or contribute toward that risk aversion."

- **At the local level, interviewees felt that the liability and career risk associated with prescribed fire is a deterrent from being more proactive with fire.** Some burners, especially with the Forest Service, said they were not always sure the agency would support them in case of an escape, whereas others felt confident that they would have legal protection from the agency as long as they acted within the scope of their duties and parameters of their burn plans. Some said they were encouraged to hold private insurance; others said this was not necessary. Another challenge may rest with the different liability laws across states; as one interviewee stated "Anytime you do prescribed fire, different states have different liability laws. Some are vague … Some are limited liability or simple negligence … there's gross negligence, simple negligence and strict liability." Several people noted that because the incentives to burn are few and hurdles to burning are many, if a line officer or FMO is more risk-averse, prescribed fire activities will be minimal on that unit.
- **Among high-level decision makers, political risk aversion and other agency practices can pose major barriers to putting fire on the landscape.** If an elected official does not support prescribed fire, this can significantly limit burning, even on federal lands. One interviewee in Washington said, "[The] personality of the person that's talking to the burner, the person signing the permit, all the

way up to the Commissioner of Public Lands, who's an elected official . . . if the elected official is extremely risk-averse, that pretty much shuts down burning. If [that person] is very proactive about forest health, we can have a little bit of risk, and maybe an intrusion and learn from it." In Colorado, one person explained that there have been limitations on their prescribed fire programs statewide due to moratoriums on burning after the escaped Lower North Fork prescribed fire and during times when fires are active on the Front Range, even when burning conditions may be excellent in other parts of the state. Due to several escaped burns in the early 2000s, the BLM put in place a system of checks and balances that make it a more lengthy and difficult process to implement prescribed burns. According to a BLM interviewee, this process still exists and is in need of updating in light of improved training and practice.

Other Challenges: Burn Windows and Other Conditions on the Ground
Limited burn windows, coupled with the challenges of getting adequate capacity during those windows, are a significant barrier in some places. Even when the conditions meet the prescription, burn windows often coincide with other considerations: a high visibility fire elsewhere, a weekend with a local festival, or a time when personnel are not available. In Arizona, one interviewee said that drought conditions meant that fuels were often too dry to burn (i.e., locations were not within prescriptions for prescribed fire). In high elevation areas, we heard that burn windows are short, because fuels can be under snow or too moist, and often come at the height of wildland fire season, making it difficult to get capacity to burn.

Other landscape conditions, like fuel types and topography, can create challenges for increased prescribed burning. Discussing fuel loads, one interviewee said, "It took a long time to get into this problem, and it's probably going to take quite a long time to get out of it. I mean, we've been suppressing wildfire for over 100 years, and it's led to a larger fuel buildup." Many BLM interviewees discussed invasive cheatgrass as a prominent barrier, as the presence of cheatgrass makes use of prescribed fire particularly challenging, providing additional incentive to rely on mechanical treatments. A Forest Service employee also described the issue, saying, "Cheatgrass, of course [it's a] huge limit to prescribed fire We're actually buying mechanical equipment because we know we can't put fire on the ground." Some places said their steep topography made burning difficult, while others said topography that facilitates inversions into populated areas makes it difficult to manage potential smoke impacts.

Species conservation requirements in some locations can conflict with application of prescribed fire. BLM interviewees all agreed that a major factor impeding the agency's ability to implement prescribed burn programs are the restrictions put into place to protect sage grouse. Another example is spotted owl habitat protection in western Oregon, which impacts burning ability, and is further exacerbated by fragmented land ownership patterns, creating, as one interviewee described, "layer(s) of Swiss cheese on the [land management] map. And [then] you're just trying to burn all the little pieces in-between that happen to be mid-slope or down in the creek, [which is] not ideal." These variables around threatened and endangered species habitat also can interact with other considerations. As one person said, "I think [it's] all the different regulations on the landscape from threatened and endangered species to just . . . trying to find that perfect time where you're in prescription, the weather's right because you're in prescription, you're in the right place at the right time, so the owls and the bugs are happy and the salamanders are happy . . .And then also I think third on the list is the smoke management approval."

A few of interviewees indicated that getting through the NEPA process creates a barrier to accomplishing more prescribed fire. Some suggested that the federal agencies find greater opportunities to undertake project planning and NEPA analysis jointly. One person explained, "I think we should be looking at being able to share, do NEPA jointly and have the Forest Service take the lead and actually work on a landscape scale that includes all federal ownerships. And then we can maybe move through that process faster, and actually get more fire on the landscape on those fringe areas where we could do joint projects." A couple people expressed a desire for less NEPA requirements. One person said it was not the law that was the problem, as much as the details of decisions made through the NEPA process by interdisciplinary teams without a fire ecologist on staff; in these cases, their plans did not adequately anticipate or support prescribed fire.

III. Opportunities and Successful Strategies to Maintain and Increase Prescribed Fire Programs

Most interviewees said that they were focused on opportunities to grow prescribed fire programs through addressing capacity and resource limitations. People in most states said air quality regulation was not their biggest challenge and pointed to the importance of the strategies they have in place that allow them to work well with air quality regulators. The exception was in Oregon and Washington where people said additional work is needed to find agreement on air quality regulations and space to increase burning with state agencies. In this section we discuss the strategies interviewees said they were using to maintain and increase their use of prescribed fire (see also Table 2, page 29).

Increased collaboration among all interested parties can be crucial to finding creative solutions to accomplish more prescribed fire. For instance, a creative opportunity is in place now in California, called the Fire MOU Partnership, which is a voluntary group that involves regulators, CALFIRE, federal land managers, and NGO partners. The group is focused on improving understanding of barriers to prescribed fire and opportunities. Working groups within this partnership are examining why burning does not occur on available burn days, and whether this is due to weather, lack of capacity, poor planning, or other variables. An agency staff member in another state explained the need for increased agency collaboration, saying, "it wouldn't hurt for us to have a little more collaboration with other agencies as far as trying to get fire on the ground ... We probably don't work with other agencies as much as we probably should . . . We could probably work a little bit together to do more landscape type projects instead of we do our projects and the [other land management agency] does their projects." These types of partnership also allow groups to find creative opportunities for resource sharing. Such efforts to foster more prescribed burning have benefits beyond prescribed fire. One Forest Service interviewee explained this, stating, "because it's the working relationships during the prescribed fire season that jumps over into the suppression season, and you already know each other, and suppression goes easily because of having those relationships in fire and fuel management already."

Coordination among burners and between air quality regulators and land managers is essential to managing competition in airsheds and capitalizing on opportunities for burners with difficult burn windows and prescriptions. One example is the Montana-Idaho Airshed Group, which is run by and for major burners (federal and state land managers, and large private landowners) to coordinate burning activities and streamline communication with DEQ. The group has a system, something that people in several other states said they wanted, for inputting and tracking all burn requests online (provided by Air Sciences, Inc.). To avoid triggering air quality concerns, the group's coordinator, staffed as a rotating position among the burners, approves all burn requests, and communicates on behalf of the burners to DEQ, so that individual burn bosses do not have to. One person explained, "I think that smoke management is a genuine challenge. I think that we can work with what we have, which has been built by burners and has been iterated by burners over the last 30 years to be as unobtrusive a smoke management approval process, as we can figure out how to build [our programs]." Burners and regulators alike in Montana pointed to the fact that regulation in the state leaves some flexibility, which is valuable for finding creative solutions and promoting communication. As we discuss more below, other states also have air quality liaisons or meetings throughout the year to coordinate among burners.

Dedicated positions and processes, particularly to navigate the intersection between prescribed fire and air quality, to bridge across land management agencies, and with air regulatory agencies, are essential. Interviewees said that open communication and trust are essential to understanding each other's concerns and finding opportunities to improve practice. As one person said, "I find that for our federal partners and for me, working with private landowners, having a strong relationship with our air quality districts, like a personal relationship, has been so important to getting projects done." The Montana/Idaho Airshed Group plays this bridging role. In most states today, the Forest Service has a dedicated liaison that works directly with air quality managers to find opportunities to burn and track multiple planned burns in airsheds. The first such position was in Arizona, where for years it was cofunded by the DEQ and Forest Service, which now funds the position entirely from the National Forest System budget; people credited this model for being essential to supporting effective burn programs. On the matter of trust, one regulator said, "When you're talking about the consequences of a decision being health consequences, if you don't trust that person or think that person might not be forthcoming with the amount of information that you may need to make a good decision, then we have to default back to a more conservative decision." This person explained that FMOs who proactively communicate within their agency and with county and regulatory partners often get their burns approved with more success than those who do not embrace communication.

People also emphasized the importance of relationships among land management agencies, at both the state and federal levels. One person gave the following example: "Those working under the FMO are very integrated and [on a] first-name-basis with their [state agency] counterparts on the fire side. And there's a good rapport and cooperation between the supervisor and the unit chief ... In the areas where we've had the biggest challenge, [that] is where either one or both of those relationships are not as strong. So I think where there's a will there's a way, and when there's not a will, there's not an incentive to find a way. And it very much does come down to those, in many cases, those relationships." In several states, people said partnerships among agencies allow them to find greater opportunities to burn, often by finding opportunities to share resources.

Communication, trust, and creative public outreach also are essential among agencies and the broader public. One person noted the importance of having interagency communication strategies and using multiple partners to communicate about fire with the public, both to build a united voice and use partners that have established trust with different stakeholder groups. Another person explained, "These [agencies] entities are basically working with limited resources and 110% workload usually. So coordinating between the agencies, if it's not mandated typically falls off the plate . . . that's I think where a lot of the issues end up happening . . . that [lack of] communication . . . that translates into really mixed messages to the public. If people aren't saying what the issues are, or why we're using prescribed fire, it creates a lot of communication barriers."

People have in place some options to share resources, although these are limited and vary in their use by state. Interviewees stated that there are ways around agreement issues such as "if you just need an engine for a day or two, most folks are more than willing to say, 'Yeah, I'll send my engine over, you send yours over, we'll just kind of do a handshake;'" however, people said this was more challenging for high-cost items. In Arizona, the land management agencies are using a Joint Powers Master Agreement to support resource sharing within the state for prescribed fire. Other regional and state offices said they are working with units to coordinate agreements to create efficiencies. Other people said they were utilizing the Good Neighbor Authority or "Wyden Authority" to share resources with the states and indicated these were useful policy tools.[20] In

---

[20] The "Good Neighbor Authority" (16 U.S.C § 2113a) allows the U.S. secretaries of Agriculture and Interior to enter into cooperative agreements or contracts with states pursuant to which state agencies can perform "forest, rangeland, and watershed restoration services" (including "activities to reduce hazardous fuels") on Forest Service and BLM land. The "Wyden Authority" (16 U.S.C. §§ 1011 & 1011a) allows the departments of Agriculture and Interior to enter into "cooperative agreements" with other federal agencies, tribal, state, and local governments, and private and nonprofit entities/ landowners for the protection/restoration/enhancement of fish/ wildlife habitat "and other resources on public or

Utah, the state established the Watershed Restoration Initiative (WRI) in the Conservation and Development Division. With a focus on wildlife habitat, WRI brings together funds and proposals from state and federal agencies as well as non-profit organizations to fund priority habitat restoration projects. Interviewees indicated that this facilitates prescribed fire projects and other fuels treatments by leveraging funding from multiple entities that has the flexibility to be used across diverse land ownerships and funding years. The program is one thing that helps provide "my fuel managers a lot more room to be strategic and to jump on when a window opens."

Improved monitoring data and smoke modeling efforts within the land management agencies can provide information that will support increased burning. As one person explained about their DEQ partners, "They've recognized that some of [their air quality requirements] really don't align with meeting the goals of protecting public health. We've got some of our meteorologists that work both for the BLM and for the Forest Service . . . we're deploying them when we do prescribed fire. And we're doing much more intensive monitoring of atmosphere conditions. And we're starting to question some of the models that have been used in the past to help determine what the ventilation index is on any given day, and therefore, how much we can burn." In both California and Utah, as well, people told us that air quality regulators and land managers are working together to identify opportunities to burn more at higher elevations, even when air quality in populated areas is poor. Doing so will require improved monitoring and modeling of smoke and could present opportunities for additional burning. In a few places, people said that individual regulators within a state sometimes would allow for different levels of burning; improved data from land managers and transparency in decision-making from air quality regulators both would be useful for making decisions more consistent and evidencebased. In some states, interviewees noted that their current or anticipated hiring of a dedicated meteorologist position in the state supports their increasing reliance on meteorology to inform smoke management in the state.

---

private land" and for "the reduction of risk from natural disaster where public safety is threatened."

Several interviewees said that the land management agencies could incorporate air quality and human health considerations more effectively into their ethos. One person, said, for example "I still think land management in the Forest Service is still really lacking air quality as a resource as something this is part of our responsibility. It's vastly improved…[but] it's still lacking…. I think…until we do that, it looks to the regulators much like we're not taking this very seriously, as if air quality is not a part of the decision-making system." A suggestion was that air quality considerations and communications training be more embedded within the cadre of personnel conducting prescribed fire. Some suggested the need for dedicated prescribed fire teams for capacity reasons, and a couple people suggested that those teams could be especially trained in communicating around smoke impacts to improve practice.

Conclusions and Recommendations

Our interviews did not yield clear indications that policy change is needed at the federal level at this time, as most interviewees said there were opportunities to increase the use of prescribed fire that would not require changes to law. However, realizing these opportunities will require creative problem-solving, and a commensurate input of staff time, funding and capacity, and leadership initiative. Two areas where policy change may be warranted are in smoke management programs in Oregon and Washington, where such revisions are underway, and potentially to facilitate easier approaches to interagency resource sharing. In addition, changes to incentive structures within the Forest Service may be warranted, and it is worth exploring possible internal practices that could alleviate current capacity limitations. We offer our targeted suggestions based on this phase of our research below.

Coordination among burners and between air quality regulators and land managers is critical to maintaining and increasing the amount of prescribed burning that occurs. Our interviews indicate that there is no "silver bullet" to increasing the application of prescribed fire, and that problem-solving requires local solutions that can only be identified through interagency coordination and problem examination. We recommend other states consider whether they would benefit from a statewide airshed group or partnerships following the models of Montana/Idaho Airshed Group and the California Fire MOU Partnership, or whether their existing partnerships and forums already serve this role. Other suggestions include the following:

- Ensure air quality liaisons or smoke coordinator positions are in place and staffed in all regions, with additional state-level positions as needed.
- Support state-level groups that promote communication among burners to manage competition within airsheds; these groups benefit from online platforms for tracking burn requests and related information.
- Improve measurement of smoke generation and dispersion to allow partners to find additional space to burn while navigating air quality concerns; targeted investment in necessary measurement techniques, equipment, and trained staff/meteorologists would be valuable.

The Forest Service, and the BLM to a lesser extent, would benefit from improved internal incentives to encourage more burning. Opportunities may include:

- Ensuring that leaders prioritize prescribed fire at all levels of the agency. One example comes from Utah, where several forest supervisors have, according to one interviewee, "set a million acre challenge in the next five years to help move the [prescribed fire] program;" the challenge is endorsed by the Governor in order to "help move the culture into more action on the ground."
- Examining targets to identify additional pathways to incentivize prescribed burning. For instance, there may be opportunities to provide targets at the national and regional levels that can only be met through prescribed fire, as compared to current fuels targets that can be met through wildland fire acres-burned or mechanical removal of fuels.
- Creating rewards or additional incentives for places that have the interest and a plan in place to increase burning. Options include dedicating funding to prescribed fire, which could be allocated by Regions or by Congress; recipients could be the agencies or collaborative efforts among community partners working together with federal land managers.
- Providing training to key fire management personnel and line officers about navigating personal liability concerns so that they are comfortable responding to positive incentives to burn more.

In a time of limited capacity and declining federal budgets, the federal agencies need more efficient avenues for sharing resources. Recommendations include:

- Providing consistent guidance on agreement mechanisms and associated requirements and developing additional personnel with the expertise to enter into and manage resource sharing agreements effectively. One option may be to reorganize contracts and agreements staff so that expertise is more centralized and advice is more consistent.

- Identifying a mechanism for sharing resources and dollars for prescribed fire activities that limits requirements for agreements. One possibility is to identify whether the Forest Service or BLM could have a budget line or authorities that would allow them to order resources from multiple agencies more efficiently, with less need for interagency agreements. As all resources are the property of the federal government, many people said they wanted to see easier ways to share resources, while still maintaining accountability.

To overcome persistent capacity challenges, personnel must be available at critical times to conduct prescribed fires. We have three suggestions for consideration:

- Dedicated prescribed fire crews could be created, either within or across agencies, and utilized more extensively. These crews would not be available for wildland firefighting, except perhaps in special circumstances, and would be trained in the unique smoke management and outreach skills that are needed in conjunction with an active prescribed fire program.
- Fire personnel could be organized such that they are more easily moved from one forest to the next, depending on the need for to conduct priority burns. We suggest actively seeking ways to utilize fire personnel more nimbly throughout the year. For instance, one Forest Service region is exploring how to put individuals with prescribed burns qualifications into the Resource Ordering and Status System (ROSS)[21] to facilitate available personnel staff being shared across forests.
- Agencies could find ways to support greater involvement of non-federal personnel (The Nature Conservancy, local fire departments, etc.) on prescribed burns.

---

[21] https://famit.nwcg.gov/applications/ROSS.

There are opportunities to improve planning to support increased application of prescribed fire. We suggest requiring that teams planning fuels reduction and forest restoration projects ensure they have members from both resource management and fire management. When this does not occur, projects often fail to incorporate plans and prescriptions for prescribed fire effectively. Regional and state program leads also should consider where there are opportunities to improve strategic planning to make sure the planning is completed and personnel are in place to capitalize on burn windows in areas that are high priority for fuels reduction.

Changes to air quality law and associated regulations at the federal level are not a priority, according to our interviewees, at this time. Most people said this was not their biggest barrier, and everyone suggested there was room for improvement related to internal agency dynamics, providing incentives, and ensuring capacity is available. In most states, people felt these factors should be addressed before focusing on air quality regulation; the exceptions were Oregon and Washington, where more collaboration and communication is needed at the state level to identify opportunities to accomplish more fire and navigate relatively more conservative air quality regulatory processes. People also said more strict PM2.5 standards will likely pose additional challenges compared to the current state of practice, and that this issue will need ongoing attention.

Among the major challenges moving forward will be finding opportunities to increase the spatial scale of burning. Landscape-level burning will generally require:

- Greater resource sharing both between agencies and other partnering organizations;
- Better engagement of private landowners, which in some places may require that the states address liability concerns for private burners; and,
- Identifying ways, given the need for such burns to last multiple days, to create flexibility with regard to air quality regulation.

Table 2. State-by-state summary of primary challenges and opportunities

	Primary reported barriers and challenges	Facilitators and opportunities	Interagency relationships for burning and air quality oversight
Arizona	• Many air-quality sensitive populations • Limited personnel capacity; resources on wildland fire • Utilizing agreement/funding mechanisms • BLM funding redirected to states with sage grouse • Non-attainment for PM 2.5 around Phoenix and Tucson • Intermixed landscape across private/federal/state lands	• 4FRI[22] is a motivator for increased Rx fire in the state • Agreements and partnerships across agencies and organizations to move resources and increase capacity. This includes the communities-at-risk agreement between BLM and State to administer private land projects • Extensive interagency communication has identified greater opportunities to burn	• Joint Powers Master Agreement allows exchange of resources across boundaries outlines joint procedures/policies • Working groups for individual counties • Arizona Conservation Partnership brings agencies together to identify priority areas based on their goals and objectives • USFS air quality liaison with DEQ in AZ • Rx fire[23] council active to support burning

---

[22] "Four Forest Restoration Initiative (www.fs.usda.gov/4fri).
[23] Rx fire: prescribed fire.

Table 2. (Continued)

	Primary reported barriers and challenges	Facilitators and opportunities	Interagency relationships for burning and air quality oversight
California	• Non-attainment areas for PM2.5 and ozone in places with high population (e.g., San Joaquin Valley) • Competition in airsheds in terms of emissions from woodstoves, farm industry, manufacturing, cars, etc. • Qualified personnel are limited and often not available due to trainings, vacations, or being pulled to wildland fire in other parts of state (year-round fire season) • Political pressure to not burn during wildfires • Qualified personnel sometimes not available to fill BLM positions Intermixed landscape across private/ federal/state lands	• Strong communication across air quality and land managers • Innovative public outreach strategies • CAL FIRE increasing commitment to Rx fire, and partnering with USFS and the Nature Conservancy (TNC) to do more • Findings opportunities to better utilize burn days, address policy issues, and identify opportunities through MOU[24] partnership • Creating more local and strategic air quality decisions based on better monitoring, data, and communication • Potential improve Forest Service strategic planning to identify and support more opportunities	• Online PIFRS (Prescribed Fire Incident Reporting System) to track multiple burn requests and facilitate permitting • Interagency, daily smoke coordination call to consider effects and feasibility of multiple planned burns • MOU between federal land managers, environmental organizations, Cal Fire, Rx fire councils, committed to common goal of increasing Rx fire and identifying problems and solutions • Air and Land Managers group, which meets twice a year to problem solve • CA and NV Smoke and Air Council • Interagency Air and Smoke Committee dedicated to technical matters like monitoring strategies • Three Rx fire councils active to support burning

---

[24] MOU: Memorandum of Understanding.

	Primary reported barriers and challenges	Facilitators and opportunities	Interagency relationships for burning and air quality oversight
Colorado	• Lack of capacity during short burn windows (resources often out on wildland fire) • Short burn windows (fuels often under snow or too moist) • Risk aversion by land managers and political leaders, especially after Lower N. Fork fire • Challenges utilizing agreement mechanisms to share resources across agencies • Mixed land ownership along Front Range • Non-attainment zones for ozone around Denver (summer)	• Committed FMOs/burn bosses who capitalize on available opportunities to burn and communicate with regulators to maintain productive relationships • Interagency resource sharing • Group of stakeholders forming to meet annually with Air Pollution Control Division	• The Air Pollution Control Division meets biennially with burners • CO Fire Prevention and Control reviews burn plans; all agencies operate under master agreement to share resources • Colorado State Forest Service are employees of Colorado State University and cannot conduct Rx burns; they burn piles as DNR employees • Rx fire council active to support burning • Annual meetings with major burners and regulators occurring in last two years
Idaho	• Short burn windows due to weather conditions and complex topography (valleys more prone to smoke intrusions) • When burning can occur, there is competition in some airsheds among multiple burners • Lack of funding and resources to conduct burning. • Non-attainment areas are already at risk of violating air quality standards • BLM funding redirected to states with sage grouse • Public communication about Rx burning has historically been limited	• Strong interagency communication • Burning goals based on available resources • Improved understanding of burn policies and how to conduct Rx fire • Dedicated meteorologist position • Opportunities lie in building a more robust SMP[25] including more communication with the public. • Potential opportunities to increase staff for	• Montana/Idaho Airshed Group coordinates burn planning across both states among federal, state, and private burners working under a MOU; group leads coordinate and communicate with DEQ on behalf of burners during ventilation hotline period • Annual burners meeting

---

[25] SMP: Smoke management plan.

Table 2. (Continued)

	Primary reported barriers and challenges	Facilitators and opportunities	Interagency relationships for burning and air quality oversight
		burn paperwork administration, increase resources in field education, and improve interagency communication	
Montana	• USFS' focus on meeting timber targets results in more mechanical thinning than Rx burns, especially when fuels acre targets until FY 18 could be met through wildland fire events • Utilizing agreement/funding mechanisms • Non-attainment for PM 10 around Missoula • Public frustration about Rx fire when their use of woodstoves or other activities may be constrained during some seasons	• The MT/ID Airshed group facilitates communication across major and non-major burners • Some burners work closely with the airshed group to make their needs known, which helps them get approval during tight windows • Flexible regulatory structure at the state level	• Montana/Idaho Airshed Group coordinates burn planning across both states among federal, state, and private burners working under a MOU; group leads coordinate and communicate with DEQ on behalf of burners during ventilation hotline period • Annual burners meeting • State has agreements with BLM and FWS, which enables them to help federal agencies conduct pile burning
Nevada	• Limited funding and human resources, often due to being pulled to fire suppression • Short burn windows for broadcast burns due to inversions • Sage grouse and cheatgrass considerations for BLM • Rx fire still somewhat sensitive in state due to Little Valley escaped Rx fire in 2016 where homes were lost. State forestry,	• MOU between BLM and NDEP • USFS-BLM fire resource sharing agreement in place • Opportunities include increased outreach to the public and providing more burn trainings to increase capacity and skills of agency employees	• BLM and NV Division of Environmental Protection (NDEP) have an agreement for Rx burning that must be re-done every five years. BLM agrees to work with the state and follow the permitting process, and BLM agrees to provide NDEP the amount of pollution the agency produces and allows NDEP to come to their burns • The USFS in R4, specifically NV, to address personnel capacity issues, has an

	Primary reported barriers and challenges	Facilitators and opportunities	Interagency relationships for burning and air quality oversight
	• which conducted the burn, hasn't done any burning since that fire • Smoke from California limits air quality in airsheds No strong sense that a great deal more Rx fire is needed		• agreement with BLM in which USFS sets aside money in an account, and if they need to use the BLM, BLM can charge to that account and be available on a fire • In process of forming an Rx fire council
New Mexico	• Public opposition to smoke in some locations • Intermixed landscape across private/ federal/state lands • Limited personnel capacity; resources on wildland fire • BLM funding redirected to states with sage grouse • Utilizing agreement/funding mechanisms	• Interagency resource sharing • Returning Heroes Wildland Firefighter Program • Future potential to review and update air quality regulatory processes; will be key to address processes for management of natural ignitions	• Annual interagency planning meeting • USFS air quality liaison with DEQ in NM • BLM and State Land Office partner in E. NM on cross-boundary burns • Southwest Coordination Group (all federal burners) • Oil and gas partnerships in place with BLM to facilitate communication to shut off the oil lines around burns • Rx fire council active to support burning
Oregon	• Short and unpredictable burn windows due to weather • Concern about potential for smoke intrusions[26] into Smoke Sensitive Receptor Areas (SSRAs). • Non-attainment areas due to wood smoke are already at risk of violating air quality standards • Endangered and threatened species protections limit Rx fire	• Improved communication between DEQ and Oregon Department of Forestry • Partnerships with NGOs to burn (e.g., TNC, Rx Fire Council) • Opportunities with SMP revision to improve techniques, increase public outreach, revise terminology	• BLM and Region 6 USFS partner together to develop supplemental interagency guidance for Rx burning • Formal and informal part-nerships between burners augment limited agency staff for burns events and can facilitate sharing of training, technical assistance, pers-onnel, equipment, and comm-unication • Rx fire council active to support burning

---

[26] Smoke intrusion: "smoke from prescribed fire entering a designated area at unacceptable levels" (NWCG, 2012).

Table 2. (Continued)

	Primary reported barriers and challenges	Facilitators and opportunities	Interagency relationships for burning and air quality oversight
	• Lower public smoke tolerance after recent wildfires • Lack of dedicated funding for burning; USFS prioritizing wildfires and BLM prioritizing sage grouse • Historically, limited dialogue statewide about Rx burning and public health tradeoffs	• Opportunities for greater investment (people and funding) in certain regions could increase Rx fire • Opportunities to bring forestry and public health experts together to create and revise relevant policy	
Utah	• Single clearing index across entire state (500 or above within 50 miles of sensitive areas) is limiting as it doesn't allow elevational and geographic differences. Some exceptions being allowed at 450 or above • Challenges of burning cheat grass discourages Rx fire • Lack of staff with needed Rx fire qualifications • Mechanical treatments more predictable in terms of capacity and funding than Rx fire to meet targets • Limited burn windows due to winter inversions • DEQ perception that agencies are writing burn plans that are overly complex • Perceived public aversion to smoke	• Flexible funding mechanisms through Watershed Restoration Initiative (WRI) facilitate Rx burns • Interagency smoke coordinator increases communication • FS working to address limited staff with Rx quals by improving the ability to share resources in the state (putting Rx personnel into Resource Ordering and Status System (ROSS)) • Forest Supervisors set million-acre challenge in next five years to move the program, with support from the Governor	• Interagency smoke coordinator working for federal land management agencies and state forestry • Watershed Restoration Initiative (WRI) of Utah's Conservation and Development Division. Brings funds and proposals together from state and federal agencies and nonprofits. Multi-agency teams rank, select, and allocate funding to projects that all parties consider high priority • MOU between all burning partners conducting Rx burns according to the best management practice guidelines of the SMP. Includes state, federal, and tribes as part of UT Regional Haze SIP. The MOU group meets • at least once a year to evaluate the effectiveness of the SMP

Air Quality Impacts of Wildfires 187

	Primary reported barriers and challenges	Facilitators and opportunities	Interagency relationships for burning and air quality oversight
Utah	• Significantly less Rx burning being done than at the inception of the National Fire Plan, but there appears to be little interest in doing more burning in the state	• Interagency committee working to consider how clearing index limits can be adjusted to create more local and strategic air quality decisions	
Washington	• Lack of capacity • Short burn windows due to weather conditions • Burn approvals on the day of the burn come too late to mobilize resources to burn • Topography (valleys) and concentrated populations in areas with smoke sensitive populations impacts burning • State contains five class 1 federal areas • Visibility protection in SMP restricts weekend burning • Lack of consistency in regulatory understandings between agencies and local and state level entities • Technical glitches with burn requests online • Limited public acceptance of smoke and fire	• Interagency communication improved Rx fire understanding • Forest Resiliency Burning Pilot to identify opportuneities for Rx fire • Interagency and partner resource sharing to burn • Community outreach through local fire depart-ents, Rx Fire Council • Rx fire trainings build capacity • Opportunities with SMP revision: more burn days/changing burn thresholds, earlier burn approval, improved communication	• BLM and Region 6 USFS partner together to develop supplemental interagency guidance for Rx burning • Rx fire council active to support burning
Wyoming	• Unpredictable weather and inversions • Non-attainment zones for PM 2.5 around Sheridan and Ozone around the Upper Green River • Sage grouse-related restrictions for BLM	• Strong interagency resource sharing of equipment to help increase capacity • Opportunities may lie in finding options in sage grouse habitat, and in creating a web-based program to document burns	• DEQ holds an annual smoke management meeting to discuss burn requirements and provide an overview of the burn program • BLM has agreements with USFS and US-FWS to share equipment on Rx burns • Land management agencies partner with NGOs to conduct Rx burns: Rocky

Table 2. (Continued)

	Primary reported barriers and challenges	Facilitators and opportunities	Interagency relationships for burning and air quality oversight
Wyoming			• Mountain Elk Foundation, Mule Deer Foundation, Trout Unlimited, Wyoming Wildlife and Natural Resource Trust

Bibliography

Cleaves, D.A., Martinez, J., and Haines, T.K. 2000. "Influences of prescribed burning activity and costs in the National Forest System." General Technical Report SRS-37. Washington, DC: USDA Forest Service.

Domike, J. R., and Zacaroli, A. C. 2016. The Clean Air Handbook. American Bar Association, Section of Environment, Energy, and Resources.

Engel K.H. 2013. "Perverse incentives: the case of wildfire smoke regulation." Ecology Law Quarterly 40, 623–672.

Fernandez, S., and Rainey, H.G. 2006. "Managing successful organizational change in the public sector." Public Administration Review 66: 168-176.

Moseley C., and Charnley S. 2014. "Understanding microprocesses of institutionalization: stewardship contracting and national forest management." Policy Sciences 47: 69-98.

NWCG [National Wildfire Coordinating Group]. 2012. "Glossary of wildland fire terminology, PMS 205." National Wildfire Coordinating Group, Boise, ID.

NIFC [National Interagency Fire Center]. 2015. "Fire Information Statistics." Available at: https://www.nifc.gov/fireInfo/fireInfo_statistics.html. Last accessed Nov. 5, 2015.

North, M., Collin, B.M., and Stephens, S. 2012. "Using fire to increase the scale, benefits, and future maintenance of fuels treatments." Journal of Forestry 110: 392-401.

Quinn-Davidson, L., and Varner, J.M. 2012. "Impediments to prescribed fire across agency, landscape and manager: an example from northern California." International Journal of Wildland Fire 21: 210-218.

Ryan, K.C., Knapp, E.E., and Varner, J.M. 2013. "Prescribed fire in North American forests and woodlands: history, current practice, and challenges." Frontiers in Ecology and the Environment 11 (online issue 1): e15-e24.

Schoennagel, T., Balck, J.K., Brenkery-Smith, H., Dennison, P.E., Harvey, B.J., and others. 2017. "Adapt to more wildfire in western North American forests as climate changes." Proceedings of the National Academy of Sciences 114: 4582-4590.

USDA and USDI [US Department of Agriculture and US Department of Interior]. 2014. 2014 Quadrennial Fire Review. Washington, DC: USDA and USDI.

IMPACT OF ANTHROPOGENIC CLIMATE CHANGE ON WILDFIRE ACROSS WESTERN US FORESTS

John T. Abatzoglou[1],* and A. Park Williams[2]

[1]Department of Geography, University of Idaho, Moscow, ID 83844
[2]Lamont–Doherty Earth Observatory, Columbia University, Palisades, NY 10964
Edited by Monica G. Turner, University of Wisconsin–Madison, Madison, WI, and approved July 28, 2016
(received for review May 5, 2016)

Abstract

Increased forest fire activity across the western continental United States (US) in recent decades has likely been enabled by a number of factors, including the legacy of fire suppression and human settlement, natural climate variability, and human-caused climate change. We use modeled climate projections to estimate the contribution of anthropogenic climate change to observed increases in eight fuel aridity metrics and forest fire area across the western United States. Anthropogenic increases in temperature and vapor pressure deficit significantly enhanced fuel aridity across western US forests over the past several decades and, during 2000–2015, contributed to 75% more forested area experiencing high (>1 σ) fire-season fuel aridity and an average of nine additional days per year of high fire potential. Anthropogenic climate change accounted for ~55% of observed increases in fuel aridity from 1979 to 2015 across western US forests, highlighting both anthropogenic climate change and natural climate variability as important contributors to increased wildfire potential in recent decades. We estimate that human-caused climate change contributed to an additional 4.2 million ha of forest fire area during 1984– 2015, nearly doubling the forest fire area expected in its absence. Natural climate variability will continue to

---

* To whom correspondence should be addressed. Email: jabatzoglou@uidaho.edu.

alternate between modulating and compounding anthropogenic increases in fuel aridity, but anthropogenic climate change has emerged as a driver of increased forest fire activity and should continue to do so while fuels are not limiting.

Keywords: wildfire, climate change, attribution, forests

Widespread increases in fire activity, including area burned (1, 2), number of large fires (3), and fire-season length (4, 5), have been documented across the western United States (US) and in other temperate and high-latitude ecosystems over the past half century (6, 7). Increased fire activity across western US forests has coincided with climatic conditions more conducive to wildfire (2–4, 8). The strong interannual correlation between forest fire activity and fire-season fuel aridity, as well as observed increases in vapor pressure deficit (VPD) (9), fire danger indices (10), and climatic water deficit (CWD) (11) over the past several decades, present a compelling argument that climate change has contributed to the recent increases in fire activity. Previous studies have implicated anthropogenic climate change (ACC) as a contributor to observed and projected increases in fire activity globally and in the western United States (12–19), yet no studies have quantified the degree to which ACC has contributed to observed increases in fire activity in western US forests.

Changes in fire activity due to climate, and ACC therein, are modulated by the co-occurrence of changes in land management and human activity that influence fuels, ignition, and suppression. The legacy of twentieth century fire suppression across western continental US forests contributed to increased fuel loads and fire potential in many locations (20, 21), potentially increasing the sensitivity of area burned to climate variability and change in recent decades (22). Climate influences wildfire potential primarily by modulating fuel abundance in fuel-limited environments, and by modulating fuel aridity in flammability-limited environments (1, 23, 24). We constrain our attention to climate processes that promote fuel aridity that encompass fire behavior characteristics of landscape ignitability, flammability, and fire spread via fuel desiccation in primarily flammability-limited western US forests by considering eight fuel aridity metrics that have well-established direct interannual relationships with burned area in this region (1, 8, 24, 25). Four metrics were calculated from monthly data for 1948–2015: (*i*) reference potential evapotranspiration (ETo), (*ii*) VPD, (*iii*) CWD, and (*iv*) Palmer drought severity index (PDSI). The other four metrics are daily fire danger indices calculated for 1979–2015: (*v*) fire weather index (FWI) from the Canadian forest fire danger rating system, (*vi*) energy release component (ERC) from the US national fire danger rating system, (*vii*) McArthur forest fire danger index (FFDI), and (*viii*) Keetch–Byram drought index (KBDI).

These metrics are further described in the *Materials and Methods* and *Supporting Information*. Fuel aridity has been a dominant driver of regional and subregional interannual variability in forest fire area across the western US in recent decades (2, 8, 22, 25). This study capitalizes on these relationships and specifically seeks to determine the portions of the observed increase in fuel aridity and area burned across western US forests attributable to anthropogenic climate change.

The interannual variability of all eight fuel aridity metrics averaged over the forested lands of the western US correlated significantly ($R^2 = 0.57$–0.76, $P < 0.0001$; Table S1) with the logarithm of annual western US forest area burned for 1984–2015, derived from the Monitoring Trends in Burn Severity product for 1984–2014 and the Moderate Resolution Imaging Spectroradiometer (MODIS) for 2015 (*Supporting Information*). The record of standardized fuel aridity averaged across the eight metrics (hereafter, all-metric mean) accounts for 76% of the variance in the burned-area record, with significant increases in both records for 1984–2015 (Figure 1). Correlation between fuel aridity and forest fire area remains highly significant ($R^2 = 0.72$, all-metric mean) after removing the linear-least squares trends for each time series for 1984–2015, supporting the mechanistic relationship between fuel aridity and forest fire area. It follows that co-occurring increases in fuel aridity and forest fire area over multiple decades would also be mechanistically related.

> **Significance**
>
> Increased forest fire activity across the western United States in recent decades has contributed to widespread forest mortality, carbon emissions, periods of degraded air quality, and substantial fire suppression expenditures. Although numerous factors aided the recent rise in fire activity, observed warming and drying have significantly increased fire-season fuel aridity, fostering a more favorable fire environment across forested systems. We demonstrate that human-caused climate change caused over half of the documented increases in fuel aridity since the 1970s and doubled the cumulative forest fire area since 1984. This analysis suggests that anthropogenic climate change will continue to chronically enhance the potential for western US forest fire activity while fuels are not limiting.

Author contributions: J.T.A. and A.P.W. designed research, performed research, contributed new reagents/analytic tools, analyzed data, and wrote the paper.

The authors declare no conflict of interest. This article is a PNAS Direct Submission.

See Commentary on page 11649.

This article contains supporting information online at www.pnas.org/lookup/suppl/doi:10.1073/pnas.1607171113/-/DCSupplemental.

We quantify the influence of ACC using the Coupled Model Intercomparison Project, Phase 5 (CMIP5) multimodel mean changes in temperature and vapor pressure following Williams et al. (26) (Figure S1; *Methods*). This approach defines the ACC signal for any given location as the multimodel mean (27 CMIP5 models) 50-y low-pass-filtered record of monthly temperature and vapor pressure anomalies relative to a 1901 baseline. Other anthropogenic effects on variables such as precipitation, wind, or solar radiation may have also contributed to changes in fuel aridity but anthropogenic contributions to these variables during our study period are less certain (22). We evaluate differences between fuel aridity metrics computed with the observational record and those computed with observations that exclude the ACC signal to determine the contribution of

ACC to fuel aridity. To exclude the ACC signal, we subtract the ACC signal from daily and monthly temperature and vapor pressure, leaving all other variables unchanged and preserving the temporal variability of observations. The contribution of ACC to changes in fuel aridity is shown for the entire western United States; however, we constrain the focus of our attribution and analysis to forested environments of the western US (Figure 1, *Inset*; *Methods*).

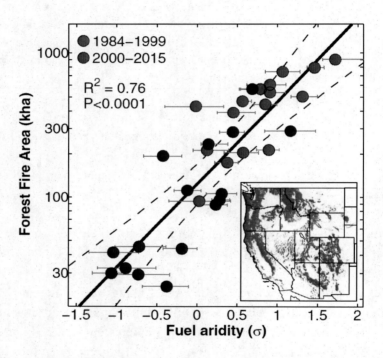

Figure 1. Annual western continental US forest fire area versus fuel aridity: 1984–2015. Regression of burned area on the mean of eight fuel aridity metrics. Gray bars bound interquartile values among the metrics. Dashed lines bounding the regression line represent 95% confidence bounds, expanded to account for lag-1 temporal autocorrelation and to bound the confidence range for the lowest correlating aridity metric. The two 16-y periods are distinguished to highlight their 3.3-fold difference in total forest fire area. *Inset* shows the distribution of forested land across the western US in green.

Anthropogenic increases in temperature and VPD contributed to a standardized (σ) increase in all-metric mean fuel aridity averaged for forested regions of +0.6 σ (range of +0.3 σ to +1.1 σ across all eight metrics) for 2000–2015 (Figure 2). We found similar results with reanalysis products (all-metric mean fuel aridity increase of +0.6 σ for two reanalysis datasets considered; *Methods*), suggesting robustness of the results to structural uncertainty in observational products (Figures S2–S4 and Table S2). The largest anthropogenic increases in standardized fuel aridity were present across the intermountain western United States, due in part to larger modeled warming rates relative to more maritime areas (27). Among aridity metrics, the largest increases tied to the ACC signal were for VPD and ETo because the interannual variability of these variables is primarily driven by temperature for much of the study area (28). By contrast, PDSI and ERC showed more subdued ACC driven increases in fuel aridity because these metrics are more heavily influenced by precipitation variability.

Fuel aridity averaged across western US forested areas showed a significant increase over the past three decades, with a linear trend of +1.2 σ (95% confidence: 0.42–2.0 σ) in the all-metric mean for 1979–2015 (Figure 3*A*, *Top* and Table S1). The all-metric mean ACC contribution since 1901 was +0.10 σ by 1979 and +0.71 σ by 2015. The annual area of forested lands with high fuel aridity (>1 σ) increased significantly during 1948–2015, most notably since 1979 (Figure 3*A*, *Bottom*). The observed mean annual areal extent of forested land with high aridity during 2000–2015 was 75% larger for the all-metric mean (+27% to +143% range across metrics) than was the case where the ACC signal was excluded.

Significant positive trends in fuel aridity for 1979–2015 across forested lands were observed for all metrics (Figure 3*B* and Table S1). Positive trends in fuel aridity remain after excluding the ACC signal, but the remaining trend was only significant for ERC. Anthropogenic forcing accounted for 55% of the observed positive trend in the all-metric mean fuel aridity during 1979– 2015, including at least two-thirds of the observed increase in ETo, VPD, and FWI, and less than a third of the observed increase in ERC and PDSI. No significant trends were observed for monthly fuel aridity metrics from 1948–1978.

The duration of the fire-weather season increased significantly across western US forests (+41%, 26 d for the all-metric mean) during 1979–2015, similar to prior results (10) (Figure 4A and Table S2). Our analysis shows that ACC accounts for ~54% of the increase in fire-weather season length in the all-metric mean (15–79% for individual metrics). An increase of 17.0 d per year of high fire potential was observed for 1979–2015 in the all-metric mean (11.7–28.4 d increase for individual metrics), over twice the rate of increase calculated from metrics that excluded the ACC signal (Figure 4B and Table S2). This translates to an average of an additional 9 d (7.8–12.0 d) per year of high fire potential during 2000–2015 due to ACC.

Given the strong relationship between fuel aridity and annual western US forest fire area, and the detectable impact of ACC on fuel aridity, we use the regression relationship in Figure 1 to model the contribution of ACC on western US forest fire area for the past three decades (Figure 5 and Figure S5). ACC-driven increases in fuel aridity are estimated to have added ~4.2 million ha (95% confidence: 2.7–6.5 million ha) of western US forest fire area during 1984–2015, similar to the combined areas of Massachusetts and Connecticut, accounting for nearly half of the total modeled burned area derived from the all-metric mean fuel aridity. Repeating this calculation for individual fuel aridity metrics yields ACC contributions of 1.9–4.9 million ha, but most individual fuel aridity metrics had weaker correlations with burned area and thus may be less appropriate proxies for attributing burned area. The effect of the ACC forcing on fuel aridity increased during this period, contributing ~5.0 (95% confidence: 4.2–5.9) times more burned area in 2000–2015 than in 1984–1999 (Figure 5B).

During 2000–2015, the ACC-forced burned area likely exceeded the burned area expected in the absence of ACC (Figure 5B). A more conservative method that uses the relationship between detrended records of burned area and fuel aridity (2) still indicates a substantial impact of ACC on total burned area, with a 19% (95% confidence: 12–24%) reduction in the proportion of total burned area attributable to ACC (Figure S5).

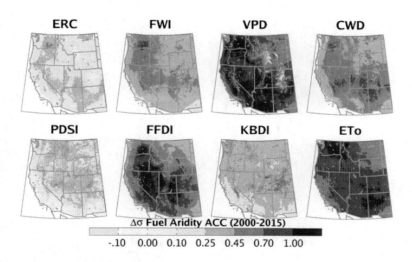

Figure 2. Standardized change in each of the eight fuel aridity metrics due to ACC. The influence of ACC on fuel aridity during 2000–2015 is shown by the difference between standardized fuel aridity metrics calculated from observations and those calculated from observations excluding the ACC signal. The sign of PDSI is reversed for consistency with other aridity measures.

Our attribution explicitly assumes that anthropogenic increases in fuel aridity are additive to the wildfire extent that would have arisen from natural climate variability during 1984–2015. Because the influence of fuel aridity on burned area is exponential, the influence of a given ACC forcing is larger in an already arid fire season such as 2012 (Figure 5A and Figure S5C). Anthropogenic increases in fuel aridity are expected to continue to have their most prominent impacts when superimposed on naturally occurring extreme climate anomalies. Although numerous studies have projected changes in burned area over the twenty-first century due to ACC, we are unaware of other studies that have attempted to quantify the contribution of ACC to recent forested burned area over the western United States. The near doubling of forested burned area we attribute to ACC exceeds changes in burned area projected by some modeling efforts to occur by the mid-twenty-first century (29, 30), but is proportionally consistent with mid-twenty-first century increases in burned area projected by other modeling efforts (17, 31–33).

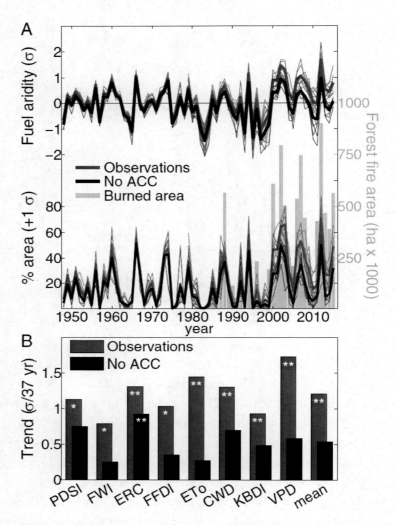

Figure 3. Evolution and trends in western US forest fuel aridity metrics over the past several decades. (*A*) Time series of (*Upper*) standardized annual fuel aridity metrics and (*Lower*) percent of forest area with standardized fuel aridity exceeding one SD. Red lines show observations and black lines show records after exclusion of the ACC signal. Only the four monthly metrics extend back to 1948. Daily fire danger indices begin in 1979. Bold lines indicate averages across fuel aridity metrics. Bars in the background of *A* show annual forested area burned during 1984–2015 for visual comparison with fuel aridity. (*B*) Linear trends in the standardized fuel aridity metrics during 1979–2015 for (red) observations and (black) records excluding the ACC signal (differences attributed to ACC). Asterisks indicate positive trends at the (*) 95% and (**) 99% significance levels.

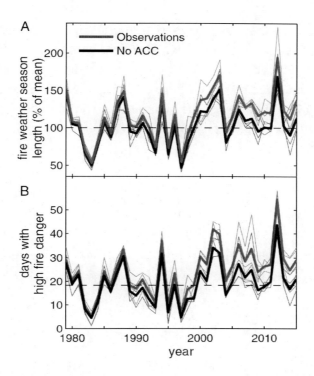

Figure 4. Changes in fire-weather season length and number of high fire danger days. Time series of mean western US forest (A) fire-weather season length and (B) number of days per year when daily fire danger indices exceeded the 95th percentile. Baseline period: 1981–2010 using observational records that exclude the ACC signal. Red lines show the observed record, and black lines show the record that excludes the ACC signal. Bold lines show the average signal expressed across fuel aridity metrics.

Beyond anthropogenic climatic changes, several additional factors have caused increases in fuel aridity and forest fire area since the 1970s. The lack of fuel aridity trends during 1948–1978 and persistence of positive trends during 1979–2015 even after removing the ACC signal implicates natural multidecadal climate variability as an important factor that buffered anthropogenic effects during 1948–1978 and compounded anthropogenic effects during 1979–2015. During 1979–2015, for example, observed Mar–Sep vapor pressure decreased significantly across many US forest areas, in marked contrast to modeled anthropogenic increases (Figure S6) (34). Significant declines in spring (Mar–May) precipitation in the southwestern United States and summer (Jun–Sep) precipitation throughout parts of the northwestern United States during 1979–2015 (Figure S7 *A* and *B*) hastened increases in fire-season fuel aridity, consistent with observed increases in the number of consecutive dry days across the region (10). Natural climate variability, including a shift toward the cold phase of the interdecadal Pacific Oscillation (35), was likely the dominant driver of observed regional precipitation trends (36) (Figure S7 *B* and *D*).

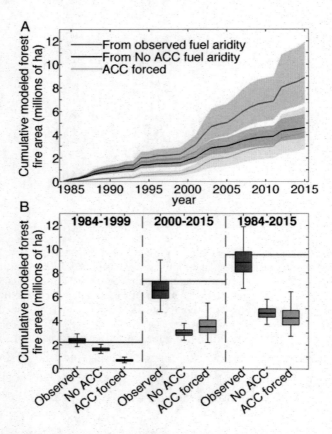

Figure 5. Attribution of western US forest fire area to ACC. Cumulative forest fire area estimated from the (red) observed all-metric mean record of fuel aridity and (black) the fuel aridity record after exclusion of ACC (No ACC). The (orange) difference is the forest fire area forced by anthropogenic increases in fuel aridity. Bold lines in *A* and horizontal lines within box plots in *B* indicate mean estimated values (regression values in Figure 1). Boxes in *B* bound 50% confidence intervals. Shaded areas in *A* and whiskers in *B* bound 95% confidence intervals. Dark red horizontal lines in *B* indicate observed forest fire area during each period.

Our quantification of the ACC contribution to observed increases in forest fire activity in the western United States adds to the limited number of climate change attribution studies on wildfire to date (37). Previous attribution efforts have been restricted to a single GCM and biophysical variable (14, 16). We complement these studies by demonstrating the influence of ACC derived from an ensemble of GCMs on several biophysical metrics that exhibit strong links to forest fire area. However, our attribution effort only considers ACC to manifest as trends in mean climate conditions, which may be conservative because climate models also project anthropogenic increases in the temporal variability of climate and drought in the western United States (34, 38, 39). In focusing exclusively on the direct impacts of ACC on fuel aridity, we do not address several other pathways by which ACC may have affected wildfire activity. For example, the fuel aridity metrics that we used may not adequately capture the role of mountain snow hydrology on soil moisture. Nor do we account for the influence of climate change on lightning activity, which may increase with warming (40). We also do not account for how fire risk may be affected by changes in biomass/fuel due to increases in atmospheric CO_2 (41), drought-induced vegetation mortality (42), or insect outbreaks (43).

Additionally, we treat the impact of ACC on fire as independent from the effects of fire management (e.g., suppression and wildland fire use policies), ignitions, land cover (e.g., exurban development), and vegetation changes beyond the degree to which they modulate the relationship between fuel aridity and forest fire area. These factors have likely added to the area burned across the western US forests and potentially amplified the sensitivity of wildfire activity to climate variability and change in recent decades (2, 22, 24, 44). Such confounding influences, along with nonlinear relationships between burned area and its drivers (e.g., Figure 1), contribute uncertainty to our empirical attribution of regional burned area to ACC. Our approach depends on the strong observed regional relationship between burned area and fuel aridity at the large regional scale of the western United States, so the quantitative results of this attribution effort are not necessarily applicable at finer spatial scales, for individual fires, or to changes in nonforested areas. Dynamical vegetation models with embedded fire models show emerging promise as tools to diagnose the impacts of a richer set of processes than those considered here (41, 45) and could be used in tandem with empirical approaches (46, 47) to better understand contributions of observed and projected ACC to changes in regional fire activity. However, dynamic models of vegetation, human activities, and fire are not without their own lengthy list of caveats (2). Given the strong empirical relationship between fuel aridity and wildfire activity identified here and in other studies (1, 2, 4, 8), and substantial increases in western US fuel aridity and fire-weather season length in recent decades, it appears clear from empirical data alone that increased fuel aridity, which is a robustly modeled result of ACC, is the proximal driver of the observed increases in western US forest fire area over the past few decades.

Conclusion

Since the 1970s, human-caused increases in temperature and vapor pressure deficit have enhanced fuel aridity across western continental US forests, accounting for approximately over half of the observed increases in fuel aridity during this period. These anthropogenic increases in fuel aridity approximately doubled the western US forest fire area beyond that expected from natural climate variability alone during 1984–2015.

The growing ACC influence on fuel aridity is projected to increasingly promote wildfire potential across western US forests in the coming decades and pose threats to ecosystems, the carbon budget, human health, and fire suppression budgets (13, 48) that will collectively encourage the development of fire-resilient landscapes (49). Although fuel limitations are likely to eventually arise due to increased fire activity (17), this process has not yet substantially disrupted the relationship between western US forest fire area and aridity. We expect anthropogenic climate change and associated increases in fuel aridity to impose an increasingly dominant and detectable effect on western US forest fire area in the coming decades while fuels remain abundant.

Methods

We focus on climate variables that directly affect fuel moisture over forested areas of the western continental United States, where fire activity tends to be flammability-limited rather than fuel- or ignition-limited (1) (study region shown in Figure 1, *Inset*). There are a variety of climate-based metrics that have been used as proxies for fuel aridity, yet there is no universally preferred metric across different vegetation types (24). We consider eight frequently used fuel aridity metrics that correlate well with fire activity variables, including annual burned area (Figure 1 and Table S1), in western US forests.

Fuel aridity metrics are calculated from daily surface meteorological data on a 1/24° grid for 1979–2015 for the western United States (west of 103°W). Although we calculated metrics across the entire western United States, we focus on forested lands defined by the climax succession vegetation stages of "forest" or "woodland" in the Environmental Site Potential product of LANDFIRE (landfire.gov). Forested 1/24° grid cells are defined by at least 50% forest coverage aggregated from LANDFIRE. We extended the aridity metrics calculated at the monthly timescale (ETo, VPD, CWD, and PDSI) back to 1948 using monthly anomalies relative to a common 1981– 2010 period from the dataset developed by the Parameterized Regression on Independent Slopes Model group (51) for temperature, precipitation, and vapor pressure, and by bilinearly interpolating NCEP–NCAR reanalysis for wind speed and surface solar radiation. We aggregated data to annualized time series of mean May–Sep daily FWI, KBDI, ERC, and FFDI; Mar–Sep VPD and ETo; Jun–Aug PDSI; and Jan–Dec CWD. We also calculated the aridity metrics strictly from ERA-INTERIM and NCEP–NCAR reanalysis products for 1979–2015 covering the satellite era (*Supporting Information*).

Days per year of high fire potential are quantified by daily fire danger indices (ERC, FWI, FFDI, and KBDI) that exceed the 95th percentile threshold defined during 1981–2010 from observations after removing the ACC signal. Observational studies have shown that fire growth preferentially occurs during high fire danger periods (52, 53). We also calculate the fire weather season length for the four daily fire danger indices following previous studies (10).

The ACC signal is obtained from ensemble members taken from 27 CMIP5 global climate models (GCMs) regridded to a common 1° resolution for 1850– 2005 using historical forcing experiments and for 2006–2099 using the Representative Concentration Pathway (RCP) 8.5 emissions scenario (Table S3 and *Supporting Information*). These GCMs were selected based on availability of monthly outputs for maximum and minimum daily temperature (T_{max} and T_{min}, respectively), specific humidity (*huss*), and surface pressure. Saturation vapor pressure (e_s), vapor pressure (*e*), and VPD were calculated using standard methods (*Supporting Information*). A variety of approaches exist to estimate the ACC signal (26). We define the anthropogenic signals in T_{max}, T_{min}, *e*, e_s, VPD, and relative humidity by a 50-y low-pass-filter time series (using a 10-point Butterworth filter) averaged across the 27 GCMs using the following methodology: For each GCM, variable, month, and grid cell, we converted each annual time series to anomalies relative toa 1901–2000 baseline. We averaged annual anomalies across all realizations (model runs) for each GCM and calculated a single 50-y low-pass-filter annual time series for each of the 12 mo for 1850–2099. We averaged each month's low-pass-filtered time series across the 27 GCMs and additively adjusted so that all smoothed records pass through zero in 1901. The resultant ACC signal represents the CMIP5 modeled anthropogenic impact since 1901 for each variable, grid cell, and month (*Supporting Information*).

We bilinearly interpolated the 1° CMIP5 multimodel mean 50-y low-pass time series to the 1/24° spatial resolution of the observations and subtracted the ACC signal from the observed daily and monthly time series. We consider the remaining records after subtraction of the ACC signal to indicate climate records that are free of anthropogenic trends (26).

Annual variations in fuel aridity metrics are presented as standardized anomalies (σ) to accommodate differences across geography and metrics. All fuel aridity metrics are standardized using the mean and SD from 1981 to 2010 for observations that excluded the ACC signal. Although the selection of a reference period can bias results (54), our findings were similar when using the full 1979–2015 time period or the observed data (without removal of ACC) for the reference period. The influence of anthropogenic forcing on fuel aridity metrics is quantified as the difference between metrics calculated with observations and those calculated with observations that excluded the ACC signal. Area-weighted standardized anomalies and the spatial extent of western US forested land that experienced high (>1 σ) aridity are computed for each aridity metric. Annualized burned area as well as aggregated fuel aridity metrics calculated with data from ref. 50 and the two reanalysis products are provided in Datasets S1–S3.

We use the regression relationship between the annual western US forest fire area and the all-metric mean fuel aridity index in Figure 1 to estimate the forcing of anthropogenic increases in fuel aridity on forest fire area during 1984–2015. Uncertainties in the regression relationship due to imperfect correlation and temporal autocorrelation are propagated as estimated confidence bounds on the anthropogenic forcing of forest fire area. This approach was repeated using a more conservative definition of the regression relationship, where we removed the linear least squares trend for 1984–2015 from both the area burned and fuel aridity time series before regression to reduce the possibility of spurious correlation due to common but unrelated trends (Figure S5). Statistical significance of all linear trends and correlations reported in this study are assessed using both Spearman's rank and Kendall's tau statistics. Trends are considered significant if both tests yield $P < 0.05$.

Acknowledgments

We thank J. Mankin, B. Osborn, and two reviewers for helpful comments on the manuscript and coauthors of ref. 26 for help developing the empirical attribution framework. A.P.W. was funded by Columbia University's Center for Climate and Life and by the Lamont-Doherty Earth Observatory (Lamont contribution 8048). J.T.A. was supported by funding from National Aeronautics and Space Administration Terrestrial Ecology Program under Award NNX14AJ14G, and the National Science Foundation

Hazards Science, Engineering and Education for Sustainability (SEES) Program under Award 1520873.

References

[1] Littell JS, McKenzie D, Peterson DL, Westerling AL (2009) Climate and wildfire area burned in western U.S. ecoprovinces, 1916-2003. *Ecol Appl* 19(4):1003–1021.

[2] Williams AP, Abatzoglou JT (2016) Recent advances and remaining uncertainties in resolving past and future climate effects on global fire activity. *Curr Clim Chang Reports* 2:1–14.

[3] Dennison P, Brewer S, Arnold J, Moritz M (2014) Large wildfire trends in the western United States, 1984–2011. *Geophys Res Lett* 41:2928–2933.

[4] Westerling AL, Hidalgo HG, Cayan DR, Swetnam TW (2006) Warming and earlier spring increase western U.S. forest wildfire activity. *Science* 313(5789):940–943.

[5] Westerling AL (2016) Increasing western US forest wildfire activity: Sensitivity to changes in the timing of spring. *Philos Trans R Soc B Biol Sci* 371(1696):20150178.

[6] Kasischke ES, Turetsky MR (2006) Recent changes in the fire regime across the North American boreal region - Spatial and temporal patterns of burning across Canada and Alaska. *Geophys Res Lett* 33(9):L09703.

[7] Kelly R, et al. (2013) Recent burning of boreal forests exceeds fire regime limits of the past 10,000 years. *Proc Natl Acad Sci USA* 110(32):13055–13060.

[8] Abatzoglou JT, Kolden CA (2013) Relationships between climate and macroscale area burned in the western United States. *Int J Wildland Fire* 22(7):1003–1020.

[9] Seager R, et al. (2015) Climatology, variability, and trends in the U.S. vapor pressure deficit, an important fire-related meteorological quantity. *J Appl Meteorol Climatol* 54(6):1121–1141.

[10] Jolly WM, et al. (2015) Climate-induced variations in global wildfire danger from 1979 to 2013. *Nat Commun* 6:7537.
[11] Dobrowski SZ, et al. (2013) The climate velocity of the contiguous United States during the 20th century. *Glob Change Biol* 19(1):241–251.
[12] Flannigan MD, Krawchuk MA, de Groot WJ, Wotton BM, Gowman LM (2009) Implications of changing climate for global wildland fire. *Int J Wildland Fire* 18(5):483–507.
[13] Flannigan M, et al. (2013) Global wildland fire season severity in the 21st century. *For Ecol Manage* 294:54–61.
[14] Yoon J, Kravitz B, Rasch P (2015) Extreme fire season in California: A glimpse into the future? *Bull Am Meteorol Soc* 96:S5–S9.
[15] Barbero R, Abatzoglou JT, Larkin NK, Kolden CA, Stocks B (2015) Climate change presents increased potential for very large fires in the contiguous United States. *Int J Wildland Fire* 24(7):892–899.
[16] Gillett NP, Weaver AJ, Zwiers FW, Flannigan MD (2004) Detecting the effect of climate change on Canadian forest fires. *Geophys Res Lett* 31(18):L18211.
[17] Westerling AL, Turner MG, Smithwick EAH, Romme WH, Ryan MG (2011) Continued warming could transform Greater Yellowstone fire regimes by mid-21st century. *Proc Natl Acad Sci USA* 108(32):13165–13170.
[18] Krawchuk MA, Moritz MA, Parisien MA, Van Dorn J, Hayhoe K (2009) Global pyrogeography: The current and future distribution of wildfire. *PLoS One* 4(4):e5102.
[19] Moritz MA, et al. (2012) Climate change and disruptions to global fire activity. *Ecosphere* 3(6):1–22.
[20] Marlon JR, et al. (2012) Long-term perspective on wildfires in the western USA. *Proc Natl Acad Sci USA* 109(9):E535–E543.
[21] Parks SA, et al. (2015) Wildland fire deficit and surplus in the western United States, 1984–2012. *Ecosphere* 6(12):1–13.
[22] Higuera PE, Abatzoglou JT, Littell JS, Morgan P (2015) The changing strength and nature of fire–climate relationships in the northern Rocky Mountains, U.S.A., 1902-2008. *PLoS One* 10(6):e0127563.

[23] Pausas JG, Ribeiro E (2013) The global fire–productivity relationship. *Glob Ecol Biogeogr* 22(6):728–736.
[24] Littell JS, Peterson DL, Riley KL, Liu Y, Luce CH (2016) A review of the relationships between drought and forest fire in the United States. *Glob Change Biol* 22(7): 2353–2369.
[25] Williams AP, et al. (2015) Correlations between components of the water balance and burned area reveal new insights for predicting forest fire area in the southwest United States. *Int J Wildland Fire* 24(1):14–26.
[26] Williams AP, et al. (2015) Contribution of anthropogenic warming to California drought during 2012–2014. *Geophys Res Lett* 42(16):6819–6828.
[27] Sheffield J, et al. (2013) North American Climate in CMIP5 experiments. Part I: Evaluation of historical simulations of continental and regional climatology. *J Clim* 26(23): 9209–9245.
[28] Hobbins MT (2016) The variability of ASCE standardized reference evapotranspiration: A rigorous, CONUS-wide decomposition and attribution. *Trans Am Soc Agric Biol Eng* 59(2):561–576.
[29] Mann ML, et al. (2016) Incorporating anthropogenic influences into fire probability models: Effects of human activity and climate change on fire activity in California. *PLoS One* 11(4):e0153589.
[30] Yue X, Mickley LJ, Logan JA, Kaplan JO (2013) Ensemble projections of wildfire activity and carbonaceous aerosol concentrations over the western United States in the mid-21st century. *Atmos Environ (1994)* 77:767–780.
[31] Pechony O, Shindell DT (2010) Driving forces of global wildfires over the past millennium and the forthcoming century. *Proc Natl Acad Sci USA* 107(45):19167–19170.
[32] Littell JS, et al. (2010) Forest ecosystems, disturbance, and climatic change in Washington State, USA. *Clim Change* 102(1-2):129–158.
[33] Rogers BM, et al. (2011) Impacts of climate change on fire regimes and carbon stocks of the U.S. Pacific Northwest. *J Geophys Res Biogeosci* 116(G3):G03037.

[34] Williams AP, et al. (2014) Causes and implications of extreme atmospheric moisture demand during the record-breaking 2011 wildfire season in the southwestern United States. *J Appl Meteorol Climatol* 53(12):2671–2684.

[35] Dong B, Dai A (2015) The influence of the Interdecadal Pacific Oscillation on temperature and precipitation over the globe. *Clim Dyn* 45(9-10):2667–2681.

[36] Deser C, Knutti R, Solomon S, Phillips AS (2012) Communication of the role of natural variability in future North American climate. *Nat Clim Chang* 2(11):775–779.

[37] National Academies of Sciences, Engineering, and Medicine (2016) *Attribution of Extreme Weather Events in the Context of Climate Change* (The National Academies Press, Washington, DC).

[38] Swain DL, Horton DE, Singh D, Diffenbaugh NS (2016) Trends in atmospheric patterns conducive to seasonal precipitation and temperature extremes in California. *Sci Adv* 2(4):e1501344.

[39] Polade SD, Pierce DW, Cayan DR, Gershunov A, Dettinger MD (2014) The key role of dry days in changing regional climate and precipitation regimes. *Sci Rep* 4:4364.

[40] Romps DM, Seeley JT, Vollaro D, Molinari J (2014) Climate change. Projected increase in lightning strikes in the United States due to global warming. *Science* 346(6211): 851–854.

[41] Knorr W, Jiang L, Arneth A (2016) Climate, CO_2 and human population impacts on global wildfire emissions. *Biogeosciences* 13(1):267–282.

[42] Williams AP, et al. (2013) Temperature as a potent driver of regional forest drought stress and tree mortality. *Nat Clim Chang* 3(3):292–297.

[43] Hart SJ, Schoennagel T, Veblen TT, Chapman TB (2015) Area burned in the western United States is unaffected by recent mountain pine beetle outbreaks. *Proc Natl Acad Sci USA* 112(14):4375–4380.

[44] Van Wagtendonk JW (2007) The history and evolution of wildland fire use. *Fire Ecol* 3(2):3–17.

[45] Bowman DMJS, Murphy BP, Williamson GJ, Cochrane MA (2014) Pyrogeographic models, feedbacks and the future of global fire regimes. *Glob Ecol Biogeogr* 23(7): 821–824.

[46] Parisien M-A, et al. (2014) An analysis of controls on fire activity in boreal Canada: Comparing models built with different temporal resolutions. *Ecol Appl* 24(6):1341–1356.

[47] Krawchuk MA, Moritz MA (2014) Burning issues: Statistical analyses of global fire data to inform assessments of environmental change. *Environmetrics* 25(6):472–481.

[48] Millar CI, Stephenson NL (2015) Temperate forest health in an era of emerging megadisturbance. *Science* 349(6250):823–826.

[49] Smith AMS, et al. (2016) The science of firescapes: Achieving fire-resilient communities. *Bioscience* 66(2):130–146.

[50] Abatzoglou JT (2013) Development of gridded surface meteorological data for ecological applications and modelling. *Int J Climatol* 33(1):121–131.

[51] Daly C, et al. (2008) Physiographically sensitive mapping of climatological temperature and precipitation across the conterminous United States. *Int J Climatol* 28(15): 2031–2064.

[52] Stavros EN, Abatzoglou J, Larkin NK, McKenzie D, Steel EA (2014) Climate and very large wildland fires in the contiguous Western USA. *Int J Wildland Fire* 23(7):899–914.

[53] Riley KL, Abatzoglou JT, Grenfell IC, Klene AE, Heinsch FA (2013) The relationship of large fire occurrence with drought and fire danger indices in the western USA, 1984– 2008: The role of temporal scale. *Int J Wildland Fire* 22(7):894–909.

[54] Sippel S, et al. (2015) Quantifying changes in climate variability and extremes: Pitfalls and their overcoming. *Geophys Res Lett* 42(22):9990–9998.

[55] Littell JS, Gwozdz RB (2011) Climatic water balance and regional fire years in the Pacific Northwest, USA: linking regional climate and fire at landscape scales. *The Landscape Ecology of Fire* (Springer, Dordrecht, The Netherlands), pp 117–139.

[56] Morton DC, et al. (2013) Satellite-based assessment of climate controls on US burned area. *Biogeosciences* 10(1):247–260.

[57] Stocks BJ, et al. (1989) Canadian forest fire danger rating system: An overview. *For Chron* 65(4):258–265.

[58] Westerling AL, Gershunov A, Brown TJ, Cayan DR, Dettinger MD (2003) Climate and wildfire in the western United States. *Bull Am Meteorol Soc* 84(5):595–604.

[59] Flannigan MD, et al. (2016) Fuel moisture sensitivity to temperature and precipitation: Climate change implications. *Clim Change* 134(1-2):59–71.

[60] Flannigan MD, Van Wagner CE (1991) Climate change and wildfire in Canada. *Can J Res* 21(1):66–72.

[61] Dowdy AJ, Mills GA, Finkele K, de Groot W (2010) Index sensitivity analysis applied to the Canadian Forest Fire Weather Index and the McArthur Forest Fire Danger Index. *Meteorol Appl* 17(3):298–312.

[62] Mitchell KE, et al. (2004) The multi-institution North American Land Data Assimilation System (NLDAS): Utilizing multiple GCIP products and partners in a continental distributed hydrological modeling system. *J Geophys Res Atmos* 109(D7):D07S90.

[63] Allen RG, Pereira LS, Raes D, Smith M (1998) Crop evapotranspiration-Guidelines for computing crop water requirements-FAO Irrigation and drainage paper 56. *FAO, Rome* 300(9):D05109.

[64] Willmott CJ, Rowe CM, Mintz Y (1985) Climatology of the terrestrial seasonal water cycle. *J Climatol* 5(6):589–606.

[65] Andrews PL, Loftsgaarden DO, Bradshaw LS (2003) Evaluation of fire danger rating indexes using logistic regression and percentile analysis. *Int J Wildland Fire* 12(2): 213–226.

[66] Cohen JE, Deeming JD (1985) The National Fire-Danger Rating System: basic equations. *Gen Tech Rep*:16.

[67] McArthur AG (1967) *Fire behaviour in eucalypt forests* (Forestry and Timber Bureau Leaflet 107).

[68] Griffiths D (1999) Improved formula for the drought factor in McArthur's Forest Fire Danger Meter. *Aust For* 62(3):202–206.
[69] Wallace JM, Hobbs PV (2006) *Atmospheric Science: An Introductory Survey* (Academic, Amsterdam), 2nd Ed.
[70] Eidenshink JC, et al. (2007) A project for monitoring trends in burn severity. *Fire Ecol* 3(1):3–21.
[71] Roy DP, Boschetti L, Justice CO, Ju J (2008) The collection 5 MODIS burned area product—Global evaluation by comparison with the MODIS active fire product. *Remote Sens Environ* 112(9):3690–3707.
[72] van Vuuren DP, et al. (2011) The representative concentration pathways: An overview. *Clim Change* 109(1):5–31.

FUTURE FIRE IMPACTS ON SMOKE CONCENTRATIONS, VISIBILITY, AND HEALTH IN THE CONTIGUOUS UNITED STATES

Ford[1], M. Val Martin[2], S. E. Zelasky[3], E. V. Fischer[1], S. C. Anenberg[4], L. Heald[5,6] and J. R. Pierce[1]

[1] Department of Atmospheric Science, Colorado State University, Fort Collins, CO, US

[2] Leverhulme Centre for Climate Change Mitigation, Department of Animal and Plant Sciences, University of Sheffield, Sheffield, UK

[3] Department of Environmental Sciences and Engineering, University of North Carolina at Chapel Hill, Chapel Hill, NC, US

[4] Department of Environmental and Occupational Health, The George Washington University, Washington, DC, US

[5] Department of Civil and Environmental Engineering, Massachusetts Institute of Technology, Cambridge, MA, US

[6] Department of Earth, Atmospheric and Planetary Sciences, Massachusetts Institute of Technology, Cambridge, MA, US

RESEARCH ARTICLE
10.1029/2018GH000144

Key Points:
- We provide the first estimates of future smoke health and visibility impacts in the contiguous United States using a prognostic land-fire model
- Average visibility will improve across the contiguous United States, but fire PM will reduce visibility on the worst days in western and southeastern U.S. regions
- The number of deaths attributable to total $PM_{2.5}$ will decrease, but the number attributable to fire-related $PM_{2.5}$ will double by late 21st century

Supporting Information:
- Supporting Information S1

Correspondence to:
B. Ford, bonne@atmos.colostate.edu

Citation:
Ford, B., Val Martin, M., Zelasky, S. E., Fischer, E. V., Anenberg, S. C., Heald, C. L., & Pierce, J. R. (2018). Future fire impacts on smoke concentrations, visibility, and health in the contiguous United States. *GeoHealth*, 2. https://doi.org/10.1029/ 2018GH000144

Received 24 APR 2018
Accepted 27 JUN 2018
Accepted article online 6 JUL 2018

Abstract

Fine particulate matter ($PM_{2.5}$) from U.S. anthropogenic sources is decreasing. However, previous studies have predicted that $PM_{2.5}$ emissions from wildfires will increase in the midcentury to next century, potentially offsetting improvements gained by continued reductions in anthropogenic emissions. Therefore, some regions could experience worse air quality, degraded visibility, and increases in population-level exposure. We use global climate model simulations to estimate the impacts of changing fire emissions on air quality, visibility, and premature deaths in the middle and late 21st century. We find that $PM_{2.5}$ concentrations will decrease overall in the contiguous United States (CONUS) due to decreasing anthropogenic emissions (total $PM_{2.5}$ decreases by 3% in Representative Concentration

Pathway [RCP] 8.5 and 34% in RCP4.5 by 2100), but increasing fire-related $PM_{2.5}$ (fire-related $PM_{2.5}$ increases by 55% in RCP4.5 and 190% in RCP8.5 by 2100) offsets these benefits and causes increases in total $PM_{2.5}$ in some regions.

We predict that the average visibility will improve across the CONUS, but fire-related $PM_{2.5}$ will reduce visibility on the worst days in western and southeastern U.S. regions. We estimate that the number of deaths attributable to total $PM_{2.5}$ will decrease in both the RCP4.5 and RCP8.5 scenarios (from 6% to 4–5%), but the absolute number of premature deaths attributable to fire-related $PM_{2.5}$ will double compared to early 21st century. We provide the first estimates of future smoke health and visibility impacts using a prognostic land-fire model. Our results suggest the importance of using realistic fire emissions in future air quality projections.

1. Introduction

Exposure to particulate matter ($PM_{2.5}$, particles with an aerodynamic diameter smaller than 2.5 μm) is associated with many negative health impacts (Crouse et al., 2012; Krewski et al., 2009; C. Arden Pope, 2007; C Arden Pope 3rd & Dockery, 2006), visibility degradation, and ecosystem impacts. There are many different sources of $PM_{2.5}$, both from human and natural sources. Because of the known detrimental effects of air pollution, the United States has sought to improve air quality through regulation of anthropogenic emissions. This has led to $PM_{2.5}$ improvements in most regions of the United States (e.g., Hand et al., 2013; Malm et al., 2017; U.S. Environmental Protection Agency (EPA), 2012). These $PM_{2.5}$ improvements are predicted to increase in the future with a further decrease in anthropogenic emissions (e.g., Lam et al., 2011; Leibensperger et al., 2012; Val Martin et al., 2015).

Wildfires are a large source of $PM_{2.5}$ in the United States, and studies have shown that the number of large wildfires has been increasing in the western United States due to warmer temperatures, earlier spring snowmelt, and longer fire seasons (e.g., Westerling, 2016; Westerling et al., 2006). Several studies have suggested that this trend will continue throughout the 21st century and that smoke could become the dominant source of $PM_{2.5}$ in the western United States during the fire season (e.g., (Liu et al., 2016; Yue et al., 2013). However, estimating future fire emissions and their impact on air quality is challenging. Fire trends are influenced not only by the changing climate but also by land use changes, land management choices, and human interactions (in terms of both ignition and suppression; e.g., Balch et al., 2017; Fusco et al., 2016; Prestemon et al., 2013). Most studies that have estimated future fires (risk or area burned) have relied on statistical regressions of current-day meteorological values (such as precipitation, relative humidity, and temperature) and fire indices (Liu et al., 2016; Prestemon et al., 2016; Spracklen et al., 2009; Westerling & Bryant, 2008; Yue et al., 2014), or parameterizations built off these statistical regressions (Liu et al., 2016; Yue et al., 2013, 2014) showed that these methods are able to explain 25–65% of the variance in area burned, but the efficacy is regionally dependent. When applied to future predictions, these studies suggest increases in fire emissions, specifically in the western United States, leading to increases in surface fire-related PM concentrations (Spracklen et al., 2009; Yue et al., 2013), visibility degradation (Spracklen et al., 2009), and smoke-exposure events (Liu et al., 2016). Val Martin et al. (2015) previously used Community Earth System Model (CESM) to simulate future PM concentrations in the United States with regard to changing emissions, land use, and climate. For fire emissions, they used the spatial distributions from the Representative Concentration Pathway (RCP) scenarios and then homogeneously scaled the monthly emissions in the western United States and Canada to match the total fire emissions from Yue et al. (2013).

More recently, process-based fire modules embedded in global land models have been used to estimate future fires and emissions (e.g., Knorr et al., 2017; Pierce et al., 2017). These fire modules use information on both climatic and socio-economic drivers (e.g., soil moisture, temperature, gross domestic product, and population density) to estimate area burned and fire emissions (Knorr et al., 2014; Li et al., 2013; Pierce et al., 2017). In contrast to statistical models, processed-based fire models can better represent feedbacks between emissions and climate and land use (Li et al., 2012, 2013, 2017). Additionally, they do not have to assume that the statistical relationships determined from current day observations will stay the same in the future under different climate scenarios nor do we need to either assume a statistical relationship holds for all regions or create a statistical relationship for each region. Finally, using the process-based fire module within a global model allows us to account for changes in fire emissions outside the study domain (contiguous United States) that can also impact air quality within our study domain. In this study, we use simulated concentrations of $PM_{2.5}$ generated by the CESM for early 21st century ("2000," the average of 2001–2010), midcentury ("2050," average of 2041–2050), and late 21st century ("2100," average of 2091–2099) described in Pierce et al. (2017) to estimate changes in $PM_{2.5}$ concentrations, population-level exposure, health effects, and visibility in the United States.

2. Methods and Tools

2.1. Model Simulations of Fire Emissions and Atmospheric Concentrations

We use the CESM to simulate surface-level $PM_{2.5}$ concentrations. A description and evaluation of the model is given in Tilmes et al. (2015). The model is run at 0.9° × 1.25° horizontal resolution for three periods: early 21st century (2000–2010), midcentury (2040–2050), and late century (2090–2099). Results are shown as 10-year averages (with the first year excluded for model spin-up). Ten-year time periods were run to represent climatological averages and account for interannual variability. The simulations were conducted in two separate steps: (1) simulating fire emissions using a land model and then (2) simulating air quality impacts using an atmospheric model.

First, emissions for landscape, agricultural, and peat fires were interactively simulated using the Community Land Model (CLM) v4.5 (Oleson et al., 2013), which accounts for changes in land cover, vegetation, climate change, and population (Pierce et al., 2017). These runs were conducted globally at 0.9° × 1.25° resolution for 1850 to 2100. Future fire simulations (2006–2100) were driven by monthly meteorological fields from archived CESM1 simulations with the RCP4.5 and RCP8.5 scenarios and population projections from the Shared Socioeconomic Pathways (SSPs;

Jones & O'Neill, 2016); the transition period (1850–2005) was forced with assimilated atmospheric data from the Climatic Research Unit of the National Centers for Environmental Prediction (CRUNCEP) and population data from the History Database of the Global Environment (HYDE). The transient run started from an 1850 equilibrium (spin-up) state of CLM4.5 with the fire module. Description of the original fire module and comparison with fire emission inventories are given in Li et al. (2012), and updates to the module and further validation are in Li et al. (2013) and Li et al. (2017). Li et al. (2013) and Li et al. (2017) found that CESM with the updated fire module is able to simulate the spatial distribution of fires, total area burned, fire seasonality, fire interannual variability and trends, and fire carbon emissions reasonably well compared to observations.

Second, the Community Atmospheric Model v4 fully coupled with an interactive gas-aerosol scheme (CAM-Chem) was used to simulate air quality impacts (Lamarque et al., 2012) using the fire emissions from the CLM. Meteorology in our present-day CAM-Chem simulations is free-running and not assimilated for present day. Hence, specific daily meteorological conditions in the present-day simulations do not correspond to observed conditions. Population projections are taken from the SSPs (Jones & O'Neill, 2016) and are described in detail in section 2.2. Biogenic emissions were determined using the Model of Emissions of Gases and Aerosols from Nature (MEGAN v2.1; Guenther et al., 2012). For estimating future anthropogenic emissions, we used the RCP scenarios 4.5 and 8.5 (van Vuuren et al., 2011). The RCPs are four different future climate scenarios that describe trajectories for greenhouse gas concentrations. They are referred to by the associated amount of radiative forcing that would occur by 2100 compared to preindustrial times (i.e., RCP4.5 corresponds to a +4.5 W/m^2 forcing). The RCP8.5 scenario assumes continued increases in greenhouse gas concentrations throughout the 21st century due to high populations, slow income growth, high energy demand with moderate technological changes to reduce emissions, and the absence of climate change policies. In RCP8.5, methane and carbon dioxide emissions will increase throughout the century. The RCP4.5 pathway has a gradual reduction in greenhouse gas emission rates after 2050 such that radiative forcing is stabilized shortly after 2100. It assumes a shift to lower emission energy sources, enactment of climate policies, less croplands, and more forests. In RCP4.5, carbon dioxide emissions will increase to midcentury and then decline; methane emissions will have a slight decline throughout the century. Both scenarios suggest a decrease in SO_2 and NO_x emissions (although a greater decline in NO_x emissions with the RCP4.5 scenario).

Fire emissions were provided to CAM-Chem from the CLM4.5 fire simulations described above. We note that due to internal variability in climate dynamics, an ensemble of CESM simulations could potentially provide a range of potential future smoke $PM_{2.5}$ concentrations (Kay et al., 2014). While we were only able to perform a single set of simulations due to the computational complexity of the CESM simulations, future work should consider an ensemble of simulations to better capture the potential range in the projections of future smoke concentrations. A full description of the model set-up, experimental design, and model evaluation can be found in Pierce et al. (2017).

Several different atmospheric simulations were conducted for each of the three different time periods to determine the contribution of different sources and emission regions to atmospheric concentrations in the contiguous United States (CONUS, the United States without Hawaii and Alaska; Table S1).

Our baseline simulations included all emission sources, while our Fireoff simulation turned off all fire emissions, and our TransportFireOff turned off fire emissions only in Canada, Alaska, Hawaii, and Mexico (this does not include transported smoke from fire emissions on other continents). By comparing these two sensitivity tests with the baseline simulation, we can determine the contribution of all wildfire smoke and the contribution of transported smoke to total $PM_{2.5}$ concentrations.

We calculate the surface-level $PM_{2.5}$ concentration from the model output with the following equation as in Val Martin et al. (2015):

$$PM_{2.5} = SO_4 + NH_4NO_3 + BC + 1.8^*(OC) + SOA + DUST + SSLT \quad (1)$$

$PM_{2.5}$ is the combination of sulfate, ammonium nitrate, secondary organic aerosols (SOAs), fine dust (first two size bins), fine sea salt (SSLT, first two size bins), black carbon (BC), and organic carbon (OC). BC and OC are the sums of the hydrophobic and hydrophilic components, and we use 1.8 as the OM:OC ratio following Hand et al. (2012). SOA is the sum of species formed from toluene, monoterpenes, isoprene, benzene, and xylene.

2.2. Population Projections

Also, included in the CLM simulations are population projections from the SSPs (Jones & O'Neill, 2016). Following van Vuuren et al. (2011), we use SSP3 with RCP8.5 and SSP1 with RCP4.5. SSP3 is a "fragmented world" scenario where a focus on national security and borders has hindered international development (O'Neill et al., 2017). Population growth is high in developing countries and low in industrialized countries, and migration is low.

This leads to a decline in the U.S. population by 2100 (but increased global population). SSP1 is the "sustainability" pathway, where there is rapid technological development, lower energy demand (particularly with less fossil fuel dependency), increased awareness of environmental degradation, and medium-to-high economic growth. Higher education levels lead to an overall lower global population, but fast urbanization and migration increase population density in urban areas around the CONUS (O'Neill et al., 2017). The predicted changes in population density for each scenario are shown in Figure 1.

These population projection scenarios are included in the simulations discussed in section 2.1 because demographic changes can alter fire activity due to suppression and can influence where fires occur (Balch et al., 2017; Knorr et al., 2014, 2016). We also use these population projections to estimate future smoke exposure, health effects, and determine population-weighted average concentrations.

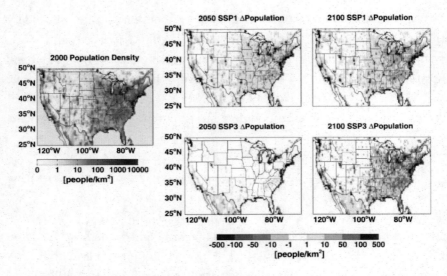

Figure 1. Figure shows the current CONUS population density (average 2006–2010) and the changes in population density projected in 2050 (average of 2040–2050) and 2100 (average of 2090–2100) by the SSP1 and SSP3 projections.

2.3. Visibility Calculations

We used the first IMPROVE equation (equation (2)) developed by Malm et al. (1994) to calculate potential changes in visibility. We chose to present results in the main text using the first equation rather than the revised equation (Pitchford et al., 2007) for better comparison with Val Martin et al. (2015).

The revised equation also separates the organic mass into large and fine mode fractions using the total mass (if the total concentration is above 20 µg/m^3, all of it is assumed to be in the large mode. If the concentration is below 20 µg/m^3, then it is separated into small and large modes, which have different mass extinction efficiencies). Using this cutoff value based on total mass to distinguish between large and small modes created some counterintuitive results when examining our sensitivity simulation results on days with high concentrations. However, we did calculate these changes in visibility using the revised IMPROVE equation and found generally similar results (see section S5).

$$b_{ext} \approx 3 \times f(RH) \times [\text{Ammonium Sulfate}] + 3 \times f(RH) \times [\text{Ammonium Nitrate}] + 4 \times [\text{Organic Mass}] + 10 \times [\text{Elemental Carbon}] + 1 \times [\text{Fine Soil}] + 0.6 \times [\text{Coarse Mass}] + \text{Rayleigh scattering}$$

(2)

With the IMPROVE equation, light extinction (b_{ext}) at each IMPROVE site is calculated by multiplying the mass concentrations (in µg/m³) of different aerosol components by typical component-specific mass extinction efficiencies. For sulfate and nitrate, the dry extinction efficiency is also multiplied by a water growth factor that is a function of relative humidity ($f(RH)$). Rayleigh scattering is assumed to be 10/Mm at every site, and gas absorption is assumed to be 0. For our results, we use the CESM ground-level daily-average relative humidity and daily-average PM species concentrations.

We also convert this to a haze index (*HI*) following equation (3) (U.S. EPA, 2003) and visual range (*VR*) following equation (4) (Pitchford & Malm, 1994) as in Val Martin et al. (2015).

$$HI = 10 * \ln(b_{ext}/10) \tag{3}$$

where b_{ext} is in inverse megameters.

$$VR = K/b_{ext} \tag{4}$$

where *K* is the Koschmieder coefficient and is assumed to be 3.91 in Pitchford and Malm (1994).

2.4. Health Impact

We calculate the all-cause mortality associated with changes in the annual-average concentrations in $PM_{2.5}$. We rely on the following concentration response function as used in Anenberg et al. (2010, 2014):

$$\Delta \text{Mortality} = \text{Pop}\left(1 - \exp^{-\beta * \Delta X}\right) Y_0 \tag{5}$$

In this equation, the change in mortality is determined by the population (Pop) and baseline mortality (Y_0) and the concentration response function. The health response (here, mortality) is related to the change in annual-mean PM$_{2.5}$ concentration (ΔX) using a concentration-response factor or beta coefficient (β) determined from relative risk (RR) estimates in epidemiological studies. The β can be determined from the RR estimate following equation (6), which is commonly used (i.e., Anenberg et al., 2010, 2014; Fann et al., 2017) and assumes a linear relationship between the ambient concentration and the log of the RR. While some studies have suggested that a linear relationship would overpredict outcomes at high concentrations (Burnett et al., 2014; Nasari et al., 2016; Pope et al., 2015), this linearity has been demonstrated over the range of PM$_{2.5}$ values relevant to this study (Krewski et al., 2009).

$$\beta = \ln(\text{RR})/\Delta X \tag{6}$$

When calculating the ΔX for equation (5), studies often subtract the threshold value (concentration below which there is no effect) or use the original epidemiological studies lowest observed concentration. For this study, we use several different β coefficients commonly used in health impact assessments in order to determine a range of estimates for all-cause mortality. To note, there are uncertainties not only in β but also in the application of the threshold/lowest-observed-concentration value, and shape of the concentration response function that will all impact our final estimates of the number of attributable premature deaths. We do investigate the impact of the threshold value for our estimates, but for a more-detailed exploration and sensitivity analyses of these uncertainties on the estimates, see Johnston et al. (2012), Kodros et al. (2018), or Ford and Heald (2016).

In this study, we use β coefficients from Krewski et al. (2009), Crouse et al. (2012), and Laden et al. (2006). The RRs, confidence intervals (CIs), and threshold/lowest-observed-concentration values from these studies are given in Table 1. To note, all of these studies are of the health effects associated with total PM$_{2.5}$ mass. Therefore, by using these β coefficients for determining the burden contribution due only to smoke, we are assuming (as has been done in other studies) that all sources and aerosol types have equal toxicity, which may not be accurate. Recent review studies specific to wildfire smoke exposure (e.g., Liu et al., 2016; Reid et al., 2016) have highlighted both similar health effects to total $PM_{2.5}$ exposure studies (positive associations with respiratory morbidity) and some distinctions (no clear association with cardiovascular morbidity). However, while there have been many studies looking at the effects of acute exposure to wildfire smoke, there are no studies that have quantified the relationship between all-cause mortality and long-term exposure to smoke $PM_{2.5}$, which is what we are determining here. Therefore, previous studies (e.g., Johnston et al., 2012) have also relied on using RR values from studies of total $PM_{2.5}$ when estimating the number of premature deaths attributable to long-term exposure to smoke $PM_{2.5}$.

Table 1. Epidemiology studies used for our calculations of attributable premature deaths with their RRs, CIs, and threshold/lowest observed concentration values

Study reference	Relative risk for $\Delta X = 10\mu g/m^3$	Confidence interval	Lowest observed/threshold concentration
Krewski et al. (2009)	1.06	1.04–1.08	5.8 µg/m³
Crouse et al. (2012)	1.10	1.05–1.15	1.9 µg/m³
Laden et al. (2006)	1.16	1.07–1.26	10 µg/m³

For our final results, we use the Krewski et al. (2009) RR because it is widely used and derived from a large cohort population in the United States (American Cancer Society Cancer Prevention Study II). We pair the Krewski et al. (2009) RR with the lowest observable concentration from Crouse et al. (2012) because subsequent research since Krewski et al. (2009) has shown mortality effects of $PM_{2.5}$ exist at $PM_{2.5}$ concentrations below the minimum observed value of Krewski et al. (2009) of 5.8 µg/m³ (Crouse et al., 2012; Pinault et al., 2016). Other studies have assumed a threshold of 0 µg/m³ (i.e., Fann et al., 2017). We also examine the sensitivity to other choices in β coefficients and threshold values, and we present those results as well.

Population is taken from the SSP1 and SSP3 projections and is on a 0.5° spatial resolution grid. For baseline mortality, we use the SSP population death rate estimates for all-cause mortality. These are given for five-year time periods for each country. We use the nationally averaged death rate for the U.S. rate for each year in our simulation period, and we regrid the $PM_{2.5}$ concentrations to the same 0.5° resolution as the population to estimate exposure concentrations. There is a decrease in the mortality rates for SSP1 and an increase for SSP3.

Finally, to attribute the number of premature deaths to each source (nonfire, CONUS fire, and AK/HI/Mexico/ Canadian transported smoke), we multiply the total number of premature deaths determined from the total $PM_{2.5}$ by the fraction of $PM_{2.5}$ from each source (determined from our sensitivity simulations). This method, as opposed to using the results from the sensitivity simulations with zeroed out emissions, avoids underestimating the contribution of sources that would occur given the dependence on the threshold value and the non-linearity of the concentration response function (Kodros et al., 2018).

3. Results

3.1. Projections of Future Smoke Emissions in North America

From the CLM, the present-day area burned for the CONUS (or Temperature North America, TENA) is 6.2 Mha/ year (for 1995–2005); this is greater than the Global Fire Emissions Database version 4 (GFED4) estimate of 1.8 Mha/year (Giglio et al., 2013); however, this GFED4 estimate does not include small fires, which are important in the United States. The GFED4s estimate, which does include small fires, estimates an average of 2.7 Mha/year for 2001–2010 with a range of ~1.5–4 Mha/year (Randerson et al., 2012). The land model simulations described in section 2.1 showed an increase in area burned in the middle and late 21st century relative to the start of the century. As the burn area increases, biomass burning (BB) emissions also increase (whereas carbon emissions from other sources are projected to decrease). The total annual average emissions of BB

BC and OC for the CONUS as determined from the CLM simulations are given in Table 2. Our early 21st century (2000) CONUS BB emissions of BC (0.058 Tg/year) are in the range of the GFED (0.011 Tg/year), Fire INventory National Center for Atmospheric Research (FINN; 0.024 Tg/year), and National Emissions Inventory (0.102 Tg/ year) inventories as given in Larkin et al. (2014). Our CONUS OC BB emissions (0.84 Tg/year) are also between the Streets et al. (2004) estimate of 0.954 Tg/year (for 1996), the U.S. EPA (2006) estimate of 0.658 Tg/year (for 2000), and FINNv1 estimate of 0.405 Tg/year (Wiedinmyer et al., 2011). Results from the CLM interactive fire simulations shown in Table 2 suggest that emissions should double by midcentury in the RCP4.5 (increase by ~50% in RCP8.5) and almost triple by 2100 in RCP8.5. As noted in Val Martin et al. (2015), the standard RCP4.5, which does not include prognostic future fire emissions from CLM (as done here) or a statistical fire prediction model (e.g., Yue et al., 2013), suggests an increase of about 60% in fire OC emissions over the western United States by 2050 while the RCP8.5 suggests a 0.3% decrease in these emissions. These assume that fire emission changes are because the standard RCP scenarios consider land use changes (afforestation in RCP4.5 and deforestation with a transition to more croplands in RCP8.5), but not any climate effects (Lamarque et al., 2010; Val Martin et al., 2015). Using a statistical fire model that did include climate changes on fires (and relied on output from 15 climate models using the A1B scenario), Yue et al. (2013) predicted a 150–170% increase in OC and BC fire emissions in the western United States by 2050. Thus, by adding in an increase in fire emissions following Yue et al. (2013), Val Martin et al. (2015) had similar increases in emissions in the western United States as shown here (~100% in RCP4.5 and ~50% in RCP8.5 increase by 2050).

Table 2. Decadal-average black carbon and organic carbon emissions in the contiguous United States due to biomass burning

CONUS biomass burning emissions	2000–2010	2040–2050		2090–2100	
	Baseline	RCP4.5	RCP8.5	RCP4.5	RCP8.5
Black carbon (Tg year)	0.058	0.13	0.087	0.12	0.14
Organic carbon (Tg year)	0.84	1.9	1.3	1.9	2.1

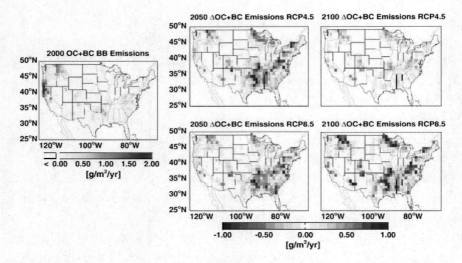

Figure 2. Early 21st century (2000), decadal average OC and BC BB emissions for the CONUS, and the changes for 2040–2050 and 2090–2100 projected with the RCP4.5 and RCP8.5 scenarios.

However, we find significant spatial differences in where the changes in BB emissions occur compared to these previous studies. In Figure 2, we show the early 21st century (average 2000–2010) BC and OC BB emissions over the CONUS and the changes for 2050 (annual average for 2040–2050) and 2100 (average 2090–2099) for the RCP4.5 and RCP8.5 scenarios determined from the land model simulations. The largest projected changes are in the southeastern United States and along the Canadian border (Figure 2). These increases in BB emissions in the eastern United States are an important distinction of this current work, as previous studies such as Yue et al. (2013) and consequently Val Martin et al. (2015) did not consider any significant increases in area burned or fire emissions over the eastern United States and only focused on the western United States. Yue et al. (2013) predicted a 150–170% increase in fire-related OC and BC emissions in the western United States by midcentury using the A1B scenario.

Here we find a 60% (RCP8.5) or 130% (RCP4.5) increase for the whole United States in midcentury; however, the majority of the increase in BC + OC emissions is for the eastern United States (85% RCP8.5, 220% RCP4.5) and not the western United States (40% RCP8.5, 45% RCP4.5). Like Yue et al. (2013), our simulations show that the western United States has peak fire emissions in August throughout the century. Additionally, the northeastern United States has a similar fire season compared to the western United States, whereas the southeastern United States has peak fire emissions earlier in May. For all regions, the annual fire emissions increase due to both increases in emissions during the peak fire season and a lengthening of the fire season, with the largest changes in both the peak emissions and lengthening occurring in the southeastern United States.

In the CLM simulations (Pierce et al., 2017), both climate and population changes drive the fire emissions. In both the RCP4.5 and RCP8.5 scenarios, the relevant climate changes (i.e., temperature, precipitation, and soil moisture) throughout the United States are overall conducive to increasing fire emissions in the future (Stocker et al., 2013; Val Martin et al., 2015, Figure S1), and climate is the main driver of our simulated changes in fire emissions in the CONUS (Pierce et al., 2017, Figure S2). The RCP4.5 scenario projects strong afforestation up to 2050 over the southeastern United States due to mitigation strategies for carbon emission reductions (shown in Val Martin et al., 2015); this rate begins to stabilize after 2050, and there is less fuel recovery by 2100. While climate is the primary driver of fire changes in the United States, population changes also impact fires, particularly the suppression and ignition of fires.

The SSP1 (used with RCP4.5) projects an increase in population over the CONUS (Figure 1), which leads to increased suppression of fires in the eastern United States in the CLM, offsetting some of the increases in fires that might be projected if only changes in climate are considered (Val Martin et al., 2015). The RCP8.5 scenario projects deforestation in much of the eastern United States and a transition to more croplands leading to less fuel available to burn.

Correspondingly, the RCP8.5 scenario does suggest a slight increase in agricultural burning in the southeast (although landscape fires overall dominate the area burned). Additionally, the SSP3 (used with RCP8.5) projects little population change by 2050 and then widespread decreases by 2100 (Figure 1). This leads to less suppression of fires, which coupled with the changes in climate, increases fire emissions significantly between the mid-century and late century.

Because our model simulations suggest that BB in the eastern United States could significantly increase, and as population and $PM_{2.5}$ concentrations are generally higher in the eastern United States compared to the western United States (with the exception of California), this could have important implications for smoke exposure and the resulting health effects.

3.2. Projections of Future PM and Fire PM in the United States

Changes in emissions will also alter $PM_{2.5}$ concentrations levels in the CONUS. By 2050, total $PM_{2.5}$ concentrations are projected to decrease primarily due to expected reductions in anthropogenic emissions in both the RCP4.5 and RCP8.5 scenarios (Figure 3). The reductions shown here are most notable in the eastern United States, particularly in the Ohio River Valley, consistent with recently observed downward trends in this region (e.g., Malm et al., 2017; U.S. EPA, 2012). In our RCP4.5 scenario simulations, $PM_{2.5}$ concentrations will continue to decrease by 2100; however, in the RCP8.5 scenario, several areas in the western United States, northeastern United States, and southeastern United States are projected to have higher concentrations compared to the early 21st century (2000). Our results shown here in Figure 3 differ from Lam et al. (2011), which showed that $PM_{2.5}$ should decrease drastically by 2050. However, they did not account for changing fire emissions.

In Figure 4, we show the decadal average of the annual changes in fire-related $PM_{2.5}$ projected in our simulations (summertime [June-July-August] average is shown in Figure S6). These results indicate that smoke concentrations are the cause for the higher $PM_{2.5}$ in our future simulations and that without an increased contribution from smoke, many regions would be projected to have even lower $PM_{2.5}$ concentrations than shown in Figure

3. Both our RCP4.5 and RCP8.5 scenarios suggest that $PM_{2.5}$ due to fire emissions will increase in the future. There are three main regions that will be impacted: (1) the Pacific Northwest (and northern California), (2) the southeastern United States, and (3) the north-central and northeastern United States along the Canadian border. In the early 21st century, fire emission accounts for more than 50% of the annual $PM_{2.5}$ only in the Pacific Northwest (Figure S6). By 2100 in both our RCP4.5 and RCP8.5 scenarios, fire emissions are projected to account for more than 50% of the annual $PM_{2.5}$ across most of the CONUS, and in areas of the three previously mentioned regions, fire emissions are projected to be responsible for 75% or more of the annual average $PM_{2.5}$ concentrations in the RCP8.5 scenario (Figure S7).

Figure 5 shows the average $PM_{2.5}$ concentrations divided by species over the CONUS from our simulations. Inorganic species (sulfate and nitrate ammonium) are predicted to decrease, while SOA and OA concentrations will increase. SOA is predicted to increase with increasing biogenic emissions as shown in Val Martin et al. (2015), while the OA increase is primarily due to fire emissions, as BB is the largest emission source of OC in the United States.

Although Val Martin et al. (2015) used different fire emissions (scaled RCP4.5 and RCP8.5 scenario fire emissions to match Yue et al., 2013), they also showed that OA concentrations would double by 2050 due to fire emissions in both the RCP4.5 and RCP8.5 scenarios. BC is predicted to decrease as mobile and industrial emissions are significant but decreasing sources, and BB is not currently the major source of BC in the United States. Yue et al. (2013) found that wildfire emissions increased western U.S. summertime OC by 46–70% and BC by 20–27% at midcentury compared to the present day.

Here we find that if we do not consider changes in anthropogenic emissions, wildfire emissions would lead to a 16% (28%) increase in summertime BC averaged for the CONUS and a 51% (86%) increase in OA in the RCP8.5 (4.5) scenario by midcentury.

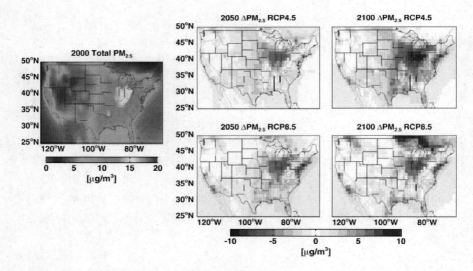

Figure 3. Total surface PM$_{2.5}$ concentrations in the CONUS for early 21st century, and the projected change (compared to early 21st century) in surface PM$_{2.5}$ concentrations by midcentury and late century from the baseline CESM simulations using the RCP4.5 and RCP8.5 scenarios.

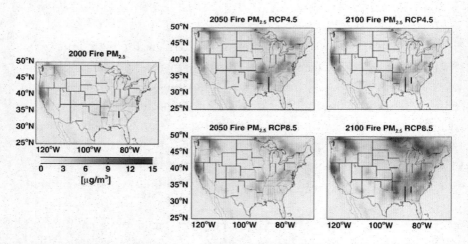

Figure 4. Simulated decadal average PM$_{2.5}$ concentrations due to fire emissions from the land model in 2000 and as projected in 2050 and 2100 in the RCP4.5 and RCP8.5 simulations (with the land model fire emissions).

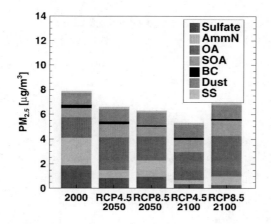

Figure 5. Average (decadal means) PM$_{2.5}$ concentrations over the CONUS separated by species for early 21st century, midcentury, and late century from the RCP4.5 and RCP8.5 scenarios.

Because of the large concentration increases along the border in the Northeast as shown in Figures 3 and 4 (where there are large population centers) projected in the RCP8.5 scenario, we also wanted to determine how much of this increase could be due to smoke transported from North American regions outside CONUS.

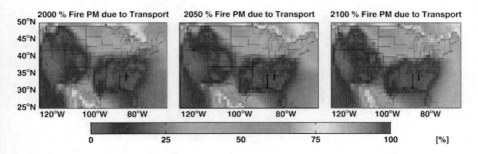

Figure 6. Percent of smoke PM$_{2.5}$ due to transport (fires outside the CONUS) for 2000 and in 2050 and 2100 with the RCP8.5 scenario.

Figure 6 shows the results from our TransportFireOff simulation (where fire emissions in Canada, Alaska, and Hawaii were turned off), which suggests that not only will concentrations increase due to local fires but also fire emissions in Canada could cause a 1–5 µg/m³ increase in PM$_{2.5}$ in the

RCP8.5 scenario (absolute concentrations are shown in Figure S8). This is approximately 50% of the smoke $PM_{2.5}$ in the northern United States, which suggests that smoke from Alaskan or Canadian fires could be responsible for 25% of the annual $PM_{2.5}$ burden in the northern United States by 2100 compared to 5% in the early 21st century.

For the CONUS-wide decadal average, the fire-related contribution to $PM_{2.5}$ concentrations is projected to go from ~25% to over 50% by 2100 in both the RCP4.5 and RCP8.5 scenarios (Figure 7; regional results are shown in Figure S9). In the RCP8.5 scenario, smoke will almost completely offset the projected reductions in nonfire $PM_{2.5}$ from the early 21st century. To note, this is for the (decadal) annual averages. During the fire season, the concentrations (and thus, the exposure concentration levels) will be even higher. In the southeast, midsouth, northeast, and west regions the summertime (June, July, and August) average in our simulations (see Figure S10) is above the World Health Organization (WHO) guideline of 10 µg/m³ and the EPA national ambient air quality standard limit of 12 µg/m³ (these standards/guidelines are for annual averages).

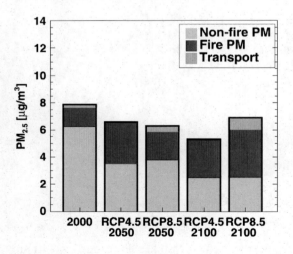

Figure 7. Decadal average $PM_{2.5}$ concentrations over the CONUS separated by source (nonfire, fire, and AK/HI/Mexico/Canadian transported smoke from fires) for early 21st century, midcentury, and late century from the RCP4.5 and RCP8.5 scenarios (simulations to determine transported smoke were only conducted for RCP8.5 scenario).

3.3. Projections of Visibility Changes in the United States

The Regional Haze Program was established with the goal of reducing visibility impairment in National Parks, forests, and historic sites around the United States. According to our simulations, visibility will increase in many of these designated areas in the eastern United States due to decreases in anthropogenic emissions. However, in the western United States where concentrations are predicted to increase due to wildland fire smoke, visibility could worsen by 2100. Additionally, wildfires are not a continuous emission source; their timing and location are sporadic, and they can produce large emission spikes on day to week timescales. Therefore, while the impact on the annual timescale may be relatively small, the contribution to a single day could be quite large. The Regional Haze Rule requires states to set goals to improve visibility and reach natural conditions on both the clearest (average of bottom 20% over 5-year period) and haziest (average of bottom 20% over 5-year period) days by 2064.

In these RCP8.5 simulations (for RCP4.5 results, see Figure S3, and for results with revised equation, see Figure S4), we see that the 20% best days are projected in our simulations to have improved visibility by 2050 and 2100 (Figure 8). However, when we look at the 20% worst days, our simulation results suggest that smoke from fires would lead to visibility degradation in many regions of western United States and the southeastern United States in 2050, which would then worsen by 2100 (Figure 8). Particular areas of vulnerability include parks in the western United States (e.g., Glacier National Park, Lassen Volcanic National Park), southeastern United States (e.g., Great Smoky Mountains National Park), and in the northeastern United States (e.g., Acadia National Park). If we compare to results from our FireOff simulation, we see that this is due to fires. Without fires, our projections suggest that visibility would continue to improve by 2050 and 2100. Visibility projections from our RCP4.5 simulations suggest similar spatial changes (visibility degradation on the worst days in the west and southeastern United States), but with different magnitudes. These results differ from the projections shown in Val Martin et al. (2015), which showed that visibility would improve on the worst and best days by 2050 in both the RCP4.5 and RCP8.5 scenarios. As we are using the same anthropogenic emissions as Val Martin et al. (2015), these differences are due to the CLM-predicted fire emissions, which have a different magnitude and spatial distribution of changes during the 21st century. As mentioned in section 3.1, Val Martin et al. (2015) only considered fire emission changes in the western United States, whereas the simulations used in this study suggest much larger changes in the eastern United States.

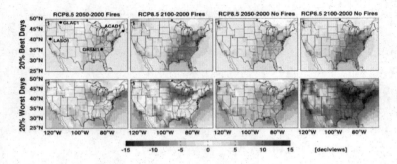

Figure 8. Change in the haze index calculated for the average of the (top row) 20% best and (bottom row) 20% worst days by 2050 and 2100 in the RCP8.5 scenario determined from our baseline simulation ("fires") and our FireOff simulation ("no fires"). Sites in Figure 9 are labeled as follows: Acadia National Park in ME (ACAD1), Great Smoky Mountains National Park in TN (GRSM1), and Lassen Volcanic National Park in northern CA (LAVO1).

In Figure 9 (Figure S4 for revised equation), we show the cumulative probability distributions of the HI at four different national park locations in the United States that will potentially experience more visibility degradation due to fires: Acadia National Park in ME (ACAD1), Glacier National Park in MT (GLAC1), Great Smoky Mountains National Park in Tennessee Mountains (GRSM1), and Lassen Volcanic National Park in northern CA (LAVO1). In general, our simulations suggest that visibility should improve in the future on the average day and on the cleanest days. At the northeastern (ACAD1) and southeastern (GRSM1) sites, fire-related PM has little impact on visibility in the early 21st century (little difference between the fire and no fire results). At the western sites, and particularly at the northern California site (LAVO1), fire-related PM has a larger impact on visibility, especially on the days with the worst visibility. For all sites, fire-related $PM_{2.5}$ will play a larger part in visibility degradation in 2050 and 2100 (in the RCP8.5 scenario) and more days will be impacted by fire-related PM compared to the early 21st century. In Figure 9, we also have the 2,064 HI targets marked for each state. Our simulation results suggest that for the four sites shown here, that smoke will make it difficult to reach the haziest day targets. Without smoke, all of the sites would be able to reach both the haziest day and clearest day targets (by 2100); with smoke, only ACAD1 will reach the clearest day goal. However, these simulation results may not be completely representative of the necessary rate of progress needed to reach the goals as we have not analyzed how well our simulations match the real baseline conditions (determined from 2000 to 2004) at each site.

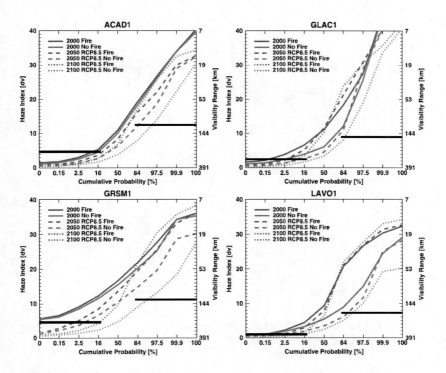

Figure 9. Cumulative probability distributions of the haze index (equation (3)) and visibility range (equation (4)) at Acadia National Park in ME (ACAD1), Great Smoky Mountains National Park in TN (GRSM1), and Lassen Volcanic National Park in northern CA (LAVO1) for our different RCP8.5 model simulations and time periods. The solid black lines show the 2064 HI targets for the clearest (average of bottom 20%) and haziest (average of top 20%) days at each site. Location of sites is noted in Figure 8.

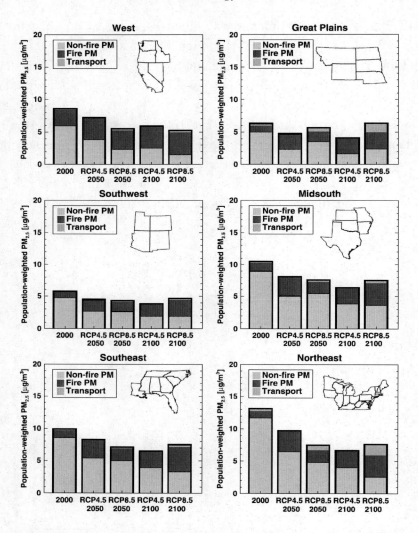

Figure 10. Decadal average of the annual population-weighted $PM_{2.5}$ concentrations for different regions of the CONUS (as defined in Val Martin et al., 2015) separated by source (nonfire, fire, and transported smoke from fires) for early 21st century (2000), midcentury (2050), and late century (2100) from the RCP4.5 and RCP8.5 scenarios (simulations to determine transported smoke were only conducted for the early 21st century and RCP8.5 projection scenarios).

3.4. Projections of Population-Level PM Exposure in the United States

While emissions and concentrations are expected to change, population is also expected to change in the future. This population change will impact the population-level exposure and the expected health effects. Thus, it is important to determine not only the average concentration but also the average exposure concentration experienced by populations in different regions.

In Figure 10, we calculate the population-weighted average concentrations for the different regions of the United States (same regions as in Val Martin et al., 2015). Both the RCP8.5 and RCP4.5 scenarios predict a decrease in the decadal average of the annual population-weighted $PM_{2.5}$ by 2050, suggesting that population-level exposure will improve for all regions of the United States. This is due both to decreasing urban emissions and population changes. However, in several regions (such as the Great Plains and Southwest), increases in fire $PM_{2.5}$ will offset a significant amount of the improvements in exposure levels associated with decreases in anthropogenic emissions. Additionally, in every region, our simulations suggest that smoke will become a dominant source of the annual average $PM_{2.5}$ exposure, even in regions that are not typically associated with wildfires, such as the Northeast and Midsouth. In these two regions, much of this increase is due to transport of smoke from other regions.

Table 3. Number (1000s) and percent (in parentheses) of premature deaths attributable to total $PM_{2.5}$ exposure per year determined from using different RRs and threshold/ lowest observed values for the different time periods and RCP scenarios

	2000	2050		2100	
Study for RR; threshold value used	Baseline	RCP4.5	RCP8.5	RCP4.5	RCP8.5
Krewski et al., 2009; $X_0 = 0.0$ μg/m³	167 (6.1%)	147 (4.8%)	145 (3.9%)	107 (3.6%)	121 (3.9%)
Krewski et al., 2009; $X_0 = 1.9$ μg/m³[a]	138 (5.1%)	114 (3.7%)	105 (2.9%)	75 (2.5%)	88 (2.9%)
Krewski et al., 2009; $X_0 = 5.8$ μg/m³	80 (2.9%)	50 (1.6%)	31 (0.84%)	17 (0.57%)	30 (0.99%)
Crouse et al., 2012; $X_0 = 1.9$ μg/m³	222 (8.1%)	185 (6.0%)	171 (4.6%)	121 (4.1%)	142 (4.7%)
Laden et al., 2006; $X_0 = 10.0$ μg/m³	69 (2.5%)	28 (0.91%)	6 (0.17%)	5 (0.17%)	16 (0.53%)

[a]Study and threshold value used for results shown in Figure 11.

In our RCP8.5 scenario simulations, population-level exposure concentrations in most regions is projected to increase or stay consistent in 2100 compared to 2050 due to increasing fire emissions offsetting decreasing nonfire emissions. For the Great Plains and Northeast, transported smoke from Canada is a significant part of this projected increase (we did not do transport sensitivity simulations for the RCP4.5 scenario).

3.5. Projections of the Health Impact of $PM_{2.5}$ and Fire $PM_{2.5}$ in the CONUS

From the $PM_{2.5}$ concentrations and population estimates, we determine the burden on premature deaths attributable to $PM_{2.5}$ exposure following the method outlined in section 2.3. In Table 3, we show results acquired using different RRs and threshold values from the studies described in the methods (Table 1). Laden et al. (2006) calculated a higher RR, but also had a higher lowest observed level, such that using this value as the threshold value causes concentrations for much of the United States to fall below this value and not contribute to the number of premature deaths attributable to $PM_{2.5}$. The highest estimates come from using the Crouse et al. (2013) RR and threshold value. The range of estimates shown here highlights the importance of the assumptions that are used in determining the health impact. Including results using these different assumptions can make our results more comparable to other studies that use different baseline assumptions.

To determine the source-specific contribution from fires, we multiplied the total premature deaths by the fraction of $PM_{2.5}$ from each source in each grid. Results are shown in Table 4 (which provides the range of estimates using the different RRs and threshold values). Both our RCP4.5 and RCP8.5 scenario simulations suggest that the number of premature deaths attributable to fire-related $PM_{2.5}$ will increase in 2050. In our simulations with the RCP4.5 scenario, it is projected that the number of attributable deaths will decrease by 2100 (but still be higher than in the early 21st century), while the number of premature deaths attributable to fire-related $PM_{2.5}$ is projected to continue to rise by 2100 in our RCP8.5 scenario simulation.

Table 4. Number (1000s) and percent (in parentheses) of premature deaths attributable to fire $PM_{2.5}$ exposure determined from using different RRs and threshold/lowest observed values for the different time periods and RCP scenarios

Study for RR; threshold value used	2000 Baseline	2050 RCP4.5	2050 RCP8.5	2100 RCP4.5	2100 RCP8.5
Krewski et al., 2009; $X_0 = 0.0$ µg/m³	21 (0.89%)	53 (1.7%)	43 (1.4%)	45 (1.5%)	59 (2.4%)
Krewski et al., 2009; $X_0 = 1.9$ µg/m³ [a]	17 (0.70%)	42 (1.4%)	32 (1.0%)	32 (1.1%)	44 (1.8%)
Krewski et al., 2009; $X_0 = 5.8$ µg/m³	10 (0.4%)	20 (0.66%)	11 (0.34%)	9 (0.31%)	18 (0.70%)
Crouse et al., 2012; $X_0 = 1.9$ µg/m³	28 (1.2%)	67 (2.2%)	51 (1.7%)	52 (1.8%)	71 (2.9%)
Laden et al., 2006; $X_0 = 10.0$ µg/m³	7 (0.31%)	15 (0.49%)	3 (0.09%)	4 (0.13%)	11 (0.41%)

[a] Study and threshold value used for results shown in Figure 11.

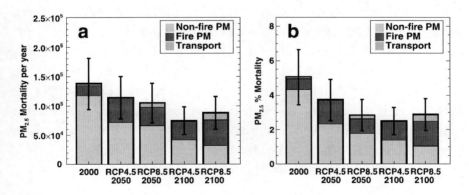

Figure 11. (a) Number and (b) percent of premature deaths attributable to $PM_{2.5}$ per year in 2000, 2050, and 2100 following the RCP4.5 and RCP8.5 scenarios and separated by source (nonfire, fire, and transported smoke). The black lines show the estimate range for the total attributable deaths from the RR CIs.

We summarize these results in Figure 11, which shows the calculated annual number (Figure 11a) and percentage (Figure 11b) of premature deaths attributable to $PM_{2.5}$ exposure in the CONUS using model-simulated $PM_{2.5}$ concentrations, the SSP population projections, the projected U.S.

mortality rates from the SSPs, the concentration-response function (equations (5) and (6)), the Krewski et al. (2009) RR listed in Table 1, and the threshold value from Crouse et al. (2013) listed in Table 1 (regional estimates are given in the Figures S11 and S12). Our results using these baseline assumptions suggest that approximately 5% of the total deaths in the CONUS are attributable to $PM_{2.5}$ in the early 21st century (range of 2-*% with different assumptions), which is in the range estimated by several previous studies (roughly 2–11% in Fann et al., 2012; Ford & Heald, 2016; Lim et al., 2012; Punger & West, 2013; and Sun et al., 2015). Estimates from our model simulations suggest that the overall number of premature deaths attributable to $PM_{2.5}$ should decrease in the United States in both the RCP4.5 and RCP8.5 scenarios. However, the U.S. population also changes throughout the century as well as the baseline mortality rates. In the SSP1 scenario (used with RCP4.5), population increases, while in the SSP3 scenario (used with RCP8.5), population declines; thus, the percent of premature deaths attributable to $PM_{2.5}$ is projected to remain at 3–4% of the total deaths. We also find (using the baseline assumptions) that 0.70% of total deaths (12.5% of the premature deaths attributable to total $PM_{2.5}$) are due to fire-related $PM_{2.5}$ in the early 21st century. The percent of deaths attributable to fire-related $PM_{2.5}$ increases by the end of the 21st century to 1.1% and 1.8% for our respective RCP4.5 and RCP8.5 cases. While the overall trends in mortality number and percent are similar in both panels of Figure 11, the qualitative differences between the two panels are due to changing population and baseline mortality rates in the future.

4. Discussion of Uncertainties

We have presented here estimates of future (2050 and 2100) air quality, health effects, and visibility in the CONUS determined from CESM simulations using emissions from a prognostic fire model. These are predictions and the veracity of the results will be limited by the model and the assumptions we made to calculate visibility and the health effects. These model simulations are uncertain due to the nature of the study which relies on RCP emission scenarios and SSPs. We are only using a single model and single simulations for each scenario. Compared to other models, CAM is more sensitive to CO_2 forcings and therefore produces stronger climate changes (Meehl et al., 2013). However, climate studies have shown that projections of decadal-mean temperatures at the end of the 21st century for specific global regions can vary greatly between simulations of even a single model due to internal variability in large-scale oceanic and atmospheric dynamics (e.g., Deser et al., 2012). Hence, we expect that an ensemble of CESM simulations would provide a range of potential future smoke $PM_{2.5}$ concentrations and associated visibility and health effects. However, due to the computational complexity of the CAM-Chem simulations at the simulated resolution, we were only able to perform one set of simulations. Future work should consider ensembles of simulations.

Additionally, the model is limited by the processes that it is able to represent. While this study does use wildfire emission estimates that were calculated with a land model (unlike many previous studies), the land model was not run dynamically/online in the CESM simulation of atmospheric concentrations. Therefore, there are potential land-atmosphere interactions that are not represented. Additionally, while the model was run at a relatively fine global resolution for a global chemistry-climate model, the grid spacing (~100 km) does not capture some important variability in concentrations relevant to the United States. At this resolution, model simulations tend to smooth out concentrations over broad regions and therefore cannot predict the high exposure concentrations often associated with dense smoke plumes or urban centers.

Not only are there large uncertainties in the $PM_{2.5}$ concentrations, but we are also limited in calculating the health burden by using the simple formulation give in equation (2). While this method is often used to provide estimates of the attributable premature deaths, studies can use different RR values, threshold values, and different formulations, or apply it to different resolutions of exposure estimates (i.e., grid level or country level), that can all lead to large uncertainties in the final numbers. A few examples were given in Table 1, but there is a large range in the RRs found in different epidemiology studies (see Ford & Heald, 2016). Additionally, we are assuming that the association between mortality and $PM_{2.5}$ remains constant over the study time period, irrespective of change in composition (we assume all $PM_{2.5}$ has the same toxicity), health care access, population activity, or other factors that might modify the relationship over time. The SSP1 and SSP3 population and mortality estimates are also model predictions that rely on many assumptions.

While there are the uncertainties described above, our results do suggest that wildfire smoke will account for a large amount of the premature deaths associated with PM exposure in the future and could offset many of the health gains from reducing anthropogenic emissions, especially by the year 2100. Future work should include ensembles of simulations and combinations of coupled land-fire-atmospheric models with statistical fire projections to quantitatively map the range of uncertainties in future projections.

5. Conclusion

This study used CESM simulations of the early 21st century, midcentury, and late century surface $PM_{2.5}$ to determine the potential impact of fires on visibility, exposure, and mortality in the CONUS. Unlike previous studies, these simulations used burn area determined from a land and fire model, which includes not only climate changes but also socioeconomic drivers. We looked at two scenarios for the future: the RCP4.5 scenario with SSP1 and the RCP8.5 scenario with SSP3 to provide the first estimates of future smoke visibility and health impacts from model simulations using emissions determined from a prognostic fire model.

Here we show, as other studies (e.g., Spracklen et al., 2009; Yue et al., 2013) have shown, that wildfire emissions will likely increase in the United States in the middle and late 21st century, while U.S. anthropogenic emissions will continue to decrease. However, unlike previous studies that focused on the western United States, our simulations suggest that there will also be significant increases in fire emissions in the southeastern United States. Our unique result could also be due to including population changes and assumptions about afforestation and deforestation in our simulations as discussed in section 3.1. Additionally, these previous studies mainly relied on parameterizations determined from statistical regressions of current day conditions while we are using a land fire model, which could explain these discrepancies. Therefore, while we are only presenting one set of simulations, these differing results do suggest that more work needs to be done using models that better account for feedbacks between climate, land use, and emissions to understand how the statistical relationships between these variables might change under different scenarios to alter fire regimes.

In many regions, the decrease in anthropogenic emissions will lead to a decrease in $PM_{2.5}$ concentrations, visibility, population-level exposure, and associated premature deaths. However, in some regions of the United States, the potential improvements will be partially offset by increases in wildfire emissions. Results from the CLM suggest that BC and OC emissions from fires will double with the largest changes in the western United States, along the Canadian border, and in the southeastern United States. By 2100, both the RCP4.5 and RCP8.5 scenarios suggest that fire-related PM will account for more than 50% of the annual average $PM_{2.5}$ concentration in the CONUS. This will be due to both local fires and transported smoke. Smoke transported from outside the CONUS (AK/HI/Mexico/Canada) could account for >50% of the fire-related PM in the Great Plains and Northeast regions in the RCP8.5 scenario.

Most of our results are for the decadal average, but wildfire smoke tends to be seasonal with large daily variability. Therefore, when looking at visibility, we saw that while the average visibility will improve, the visibility on the worst days could get even worse, particularly in the western United States, southeastern United States, and northeastern United States. We project that wildfire smoke will be the main cause of visibility degradation on the worst days in these regions.

Previous studies have quantitatively determined relationships between $PM_{2.5}$ exposure and premature deaths. Using these relationships, we calculated the burden for the early 21st century and future scenarios. We found that approximately 138,000 deaths (5.1% of total deaths) are attributable to total $PM_{2.5}$ in the early 21st century with 17,000 (0.7%) of these deaths attributable to fire-related $PM_{2.5}$. The number of total deaths attributable to $PM_{2.5}$ is projected to decrease in both scenarios over the next century, but the number attributable to fire-related PM will increase to 42,000 (1.4%, RCP4.5) or 32,000 (1.0%, RCP8.5) by 2050 and 32,000 (1.1%, RCP4.5) or 44,000 (1.8%, RCP8.5) by 2100.

Fires are potentially less controllable than urban and anthropogenic emission sources, and although there has been increased efforts to better manage fuels and forests in the United States to reduce wildfire risk, the number and intensity of wildfires has continued to increase. This is in large part due to the fact that fire frequency and intensity are strongly linked to the climate. While it is difficult to confidently determine how much the health burden could be reduced under a future climate with an RCP4.5 scenario compared to an RCP8.5 scenario (and decoupled from the changes in population) from our limited set of simulations, mitigation of climate change that could lead to a less warm and dry future climate should reduce the potential fire risks. In our simulations, we also saw that population changes had an impact on our exposure and mortality estimates, and more people are currently moving into the wildland-urban interface in the western United States, leading to a greater risk of wildfire smoke exposure. Additionally, both the RCP4.5 and RCP8.5 scenarios suggested that while the overall $PM_{2.5}$ health burden would decrease, the fraction attributable to smoke exposure could increase in the future. Therefore, to continue to reduce the health burden associated with $PM_{2.5}$ in the CONUS, more emphasis will need to be put on reducing fire-related PM exposure through public health campaigns (installing filters, creating clean air shelters, etc.) in conjunction with climate mitigation efforts.

Acknowledgments

This work was supported by the Joint Fire Science Program (grant 13-1-01-4) and the NASA Applied Sciences Program (grant NNX15AG35G). We thank Fang Li (Chinese Academy of Sciences) and David Lawrence (NCAR) for providing support with the fire module. The CESM project is supported by the National Science Foundation and the Office of Science (BER) of the U.S. Department of Energy. Computing resources were provided by the Climate Simulation Laboratory at NCAR's Computational and Information Systems Laboratory (CISL) under a Large University Computing Grant awarded to Maria Val Martin. Sarah Zelasky was supported by the National Science Foundation Research Experiences for Undergraduates Site in Climate Science at Colorado State University under the cooperative agreement AGS-1461270. Maria Val Martin was supported by the Leverhulme Trust through a Leverhulme Research Centre Award (RC-2015-029). Data available at https:// www.fs.usda.gov/rds/archive/ Product/ RDS-2018-0021 (Val Martin et al., 2018).

References

Anenberg, S. C., Horowitz, L. W., Tong, D. Q., & West, J. J. (2010). An estimate of the global burden of anthropogenic ozone and fine particulate matter on premature human mortality using atmospheric modeling. *Environmental Health Perspectives*, *118*(9), 1189–1195. https://doi. org/10.1289/ehp.0901220.

Anenberg, S. C., West, J. J., Yu, H., Chin, M., Schulz, M., Bergmann, D., et al. (2014). Impacts of intercontinental transport of anthropogenic fine particulate matter on human mortality. *Air Quality, Atmosphere and Health*, *7*(3), 369–379. https://doi.org/10.1007/s11869-014-0248-9.

Balch, J. K., Bradley, B. A., Abatzoglou, J. T., Nagy, R. C., Fusco, E. J., & Mahood, A. L. (2017). Human-started wildfires expand the fire niche across the United States. *Proceedings of the National Academy of Sciences*, *114*(11), 2946–2951. https://doi.org/10.1073/pnas.16173 94114.

Burnett, R. T., Pope, C. A. III, Ezzati, M., Olives, C., Lim, S. S., Mehta, S., et al. (2014). An integrated risk function for estimating the global burden of disease attributable to ambient fine particulate matter exposure. *Environmental Health Perspectives*. https://doi.org/10.1289/ ehp.1307049.

Crouse, D. L., Peters, P. A., van Donkelaar, A., Goldberg, M. S., Villeneuve, P. J., Brion, O., et al. (2012). Risk of nonaccidental and cardiovascular mortality in relation to long-term exposure to low concentrations of fine particulate matter: A Canadian National-Level Cohort Study. *Environmental Health Perspectives*, *120*(5), 708–714. https://doi.org/10.1289/ehp.1104049.

Deser, C., Phillips, A., Bourdette, V., & Teng, H. (2012). Uncertainty in climate change projections: The role of internal variability. *Climate Dynamics*, *38*(3–4), 527–546. https://doi.org/10.1007/s00382-010-0977-x.

Fann, N., Kim, S.-Y., Olives, C., & Sheppard, L. (2017). Estimated changes in life expectancy and adult mortality resulting from declining $PM_{2.5}$ exposures in the contiguous United States: 1980-2010. *Environmental Health Perspectives*, *125*(9), 97003. https://doi.org/10.1289/EHP507.

Fann, N., Lamson, A. D., Anenberg, S. C., Wesson, K., Risley, D., & Hubbell, B. J. (2012). Estimating the National Public Health Burden associated with exposure to ambient $PM_{2.5}$ and ozone. *Risk Analysis*, *32*(1), 81–95. https://doi.org/10.1111/j.1539-6924.2011.01630.x.

Ford, B., & Heald, C. L. (2016). Exploring the uncertainty associated with satellite-based estimates of premature mortality due to exposure to fine particulate matter. *Atmospheric Chemistry and Physics*, *16*(5), 3499–3523. https://doi.org/10.5194/acp-16-3499-2016.

Fusco, E. J., Abatzoglou Balch, J. K., Finn, J. T., & Bradley, B. A. (2016). Quantifying the human influence on fire ignition across the western USA. *Ecological Applications*, *26*(8), 2390–2401. https://doi.org/10.1002/eap.1395.

Giglio, L., Randerson, J. T., & van der Werf, G. R. (2013). Analysis of daily, monthly, and annual burned area using the fourth-generation global fire emissions database (GFED4). *Journal of Geophysical Research: Biogeosciences*, *118*, 317–328. https://doi.org/10.1002/jgrg.20042.

Guenther, A. B., Jiang, X., Heald, C. L., Sakulyanontvittaya, T., Duhl, T., Emmons, L. K., & Wang, X. (2012). The Model of Emissions of Gases and Aerosols from Nature version 2.1 (MEGAN2.1): an extended and updated framework for modeling biogenic emissions. *Geoscientific Model Development*, 5, 1471–1492. https://doi.org/10.5194/gmd-5-1471-2012.

Hand, J. L., Schichtel, B. A., Malm, W. C., & Frank, N. H. (2013). Spatial and temporal trends in PM2.5 organic and elemental carbon across the United States [research article]. Retrieved March 14, 2018, from https://www.hindawi.com/journals/amete/2013/367674/.

Hand, J. L., Schichtel, B. A., Pitchford, M., Malm, W. C., & Frank, N. H. (2012). Seasonal composition of remote and urban fine particulate matter in the United States. *Journal of Geophysical Research*, *117*, D05209. https://doi.org/10.1029/2011JD017122.

Johnston, F. H., Henderson, S. B., Chen, Y., Randerson, J. T., Marlier, M., DeFries, R. S., et al. (2012). Estimated global mortality attributable to smoke from landscape fires. *Environmental Health Perspectives*, *120*(5), 695–701. https://doi.org/10.1289/ehp.1104422.

Jones, B., & O'Neill, B. C. (2016). Spatially explicit global population scenarios consistent with the shared socioeconomic pathways. *Environmental Research Letters*, *11*(8), 10. https://doi.org/10.1088/1748-9326/11/8/084003.

Kay, J. E., Deser, C., Phillips, A., Mai, A., Hannay, C., Strand, G., et al. (2014). The Community Earth System Model (CESM) large ensemble project: A community resource for studying climate change in the presence of internal climate variability. *Bulletin of the American Meteorological Society*, *96*(8), 1333–1349. https://doi.org/10.1175/BAMS-D-13-00255.1

Knorr, W., Arneth, A., & Jiang, L. (2016). Demographic controls of future global fire risk. *Nature Climate Change*, *6*(8), 781–785. https://doi.org/10.1038/nclimate2999.

Knorr, W., Dentener, F., Lamarque, J.-F., Jiang, L., & Arneth, A. (2017). Wildfire air pollution hazard during the 21st century. *Atmospheric Chemistry and Physics*, *17*(14), 9223–9236. https://doi.org/10.5194/acp-17-9223-2017.

Knorr, W., Kaminski, T., Arneth, A., & Weber, U. (2014). Impact of human population density on fire frequency at the global scale. *Biogeosciences*, *11*(4), 1085–1102. https://doi.org/10.5194/bg-11-1085-2014.

Kodros, J. K., Carter, E., Brauer, M., Volckens, J., Bilsback, K. R., L'Orange, C., et al. (2018). Quantifying the contribution to uncertainty in mortality attributed to household, ambient, and joint exposure to $PM_{2.5}$ from residential solid fuel use. *GeoHealth*, *2*(1). 2017GH000115. doi:https://doi.org/10.1002/2017GH000115.

Krewski, D., Jerrett, M., Burnett, R. T., Ma, R., Hughes, E., Shi, Y., et al. (2009). Extended follow-up and spatial analysis of the American Cancer Society study linking particulate air pollution and mortality. *Research Report. Health Effects Institute*, *5*(140), 114–136.

Laden, F., Schwartz, J., Speizer, F. E., & Dockery, D. W. (2006). Reduction in fine particulate air pollution and mortality. *American Journal of Respiratory and Critical Care Medicine*, *173*(6), 667–672. https://doi.org/10.1164/rccm.200503-443OC.

Lam, Y. F., Fu, J. S., Wu, S., & Mickley, L. J. (2011). Impacts of future climate change and effects of biogenic emissions on surface ozone and particulate matter concentrations in the United States. *Atmospheric Chemistry and Physics*, *11*(10), 4789–4806. https://doi.org/10.5194/acp-11-4789-2011.

Lamarque, J.-F., Emmons, L. K., Hess, P. G., Kinnison, D. E., Tilmes, S., Vitt, F., et al. (2012). CAM-chem: Description and evaluation of interactive atmospheric chemistry in the community Earth system model. *Geoscientific Model Development*, *5*(2), 369–411. https://doi.org/10.5194/ gmd-5-369-2012.

Larkin, N. K., Raffuse, S. M., & Strand, T. M. (2014). Wildland fire emissions, carbon, and climate: U.S. emissions inventories. *Forest Ecology and Management*, *317*, 61–69. https://doi.org/10.1016/j.foreco.2013.09.012.

Leibensperger, E. M., Mickley, L. J., Jacob, D. J., Chen, W.-T., Seinfeld, J. H., Nenes, A., et al. (2012). Climatic effects of 1950–2050 changes in US anthropogenic aerosols—Part 1: Aerosol trends and radiative forcing. *Atmospheric Chemistry and Physics*, *12*(7), 3333–3348. https://doi. org/10.5194/acp-12-3333-2012.

Li, F., Lawrence, D. M., & Bond-Lamberty, B. (2017). Impact of fire on global land surface air temperature and energy budget for the 20th century due to changes within ecosystems. *Environmental Research Letters*, *12*(4), 44,014. https://doi.org/10.1088/1748-9326/ aa6685.

Li, F., Levis, S., & Ward, D. S. (2013). Quantifying the role of fire in the Earth system—Part 1: Improved global fire modeling in the Community Earth System Model (CESM1). *Biogeosciences*, *10*(4), 2293–2314. https://doi.org/10.5194/bg-10-2293-2013.

Li, F., Zeng, X. D., & Levis, S. (2012). A process-based fire parameterization of intermediate complexity in a dynamic global vegetation model. *Biogeosciences*, *9*(7), 2761–2780. https://doi.org/10.5194/bg-9-2761-2012.

Lim, S. S., Vos, T., Flaxman, A. D., Danaei, G., Shibuya, K., Adair-Rohani, H., et al. (2012). A comparative risk assessment of burden of disease and injury attributable to 67 risk factors and risk factor clusters in 21 regions, 1990–2010: A systematic analysis for the Global Burden of Disease Study 2010. *The Lancet*, *380*(9859), 2224–2260. https://doi.org/10.1016/S0140-6736(12)61766-8.

Liu, J. C., Mickley, L. J., Sulprizio, M. P., Dominici, F., Yue, X., Ebisu, K., et al. (2016). Particulate air pollution from wildfires in the western US under climate change. *Climatic Change*, *138*(3-4), 655–666. https://doi.org/10.1007/s10584-016-1762-6.

Malm, W. C., Schichtel, B. A., Hand, J. L., & Collett, J. L. (2017). Concurrent temporal and spatial trends in sulfate and organic mass concentrations measured in the IMPROVE monitoring program. *Journal of Geophysical Research: Atmospheres*, *122*, 10,462–10,476. https://doi.org/10.1002/ 2017JD026865.

Malm, W. C., Sisler, J. F., Huffman, D., Eldred, R. A., & Cahill, T. A. (1994). Spatial and seasonal trends in particle concentration and optical extinction in the United States. *Journal of Geophysical Research*, *99*(D1), 1347–1370. https://doi.org/10.1029/93JD02916.

Meehl, G. A., Washington, W. M., Arblaster, J. M., Hu, A., Teng, H., Kay, J. E., et al. (2013). Climate change projections in CESM1 (CAM5) compared to CCSM4. *Journal of Climate*, *26*(17), 6287–6308. https://doi.org/10.1175/JCLI-D-12-00572.1.

Nasari, M. M., Szyszkowicz, M., Chen, H., Crouse, D., Turner, M. C., Jerrett, M., et al. (2016). A class of non-linear exposure-response models suitable for health impact assessment applicable to large cohort studies of ambient air pollution. *Air Quality, Atmosphere and Health*, *9*(8), 961–972. https://doi.org/10.1007/s11869-016-0398-z..

Oleson, K. W., Lawrence, D. M., Bonan, G. B., Drewniack, B., Huang, M., Koven, C. D., Levis S., Li F., Riley W. J., Subin Z. M., Swenson S. C., Thornton P. E. (2013). Technical description of version 4.5 of the Community Land Model (CLM) (Technical Note No. NCAR/TN-503+STR). Boulder, CO: National Center for Atmospheric Research Earth System Laboratory. Retrieved from http://www.cesm.ucar.edu/models/cesm1.2/clm/CLM45_Tech_Note.pdf.

O'Neill, B. C., Kriegler, E., Ebi, K. L., Kemp-Benedict, E., Riahi, K., Rothman, D. S., et al. (2017). The roads ahead: Narratives for Shared Socioeconomic Pathways describing world futures in the 21st century. *Global Environmental Change*, *42*, 169–180. https://doi.org/10.1016/j.gloenvcha.2015.01.004.

Pierce, J. R., Val Martin, M., & Heald, C. L. (2017). Estimating the Effects of Changing Climate on Fires and Consequences for U.S. Air Quality, Using a Set of Global and Regional Climate Models (final report no. JFSP-13-1-01-4). Retrieved from https://www.firescience.gov/projects/13-1-01-4/project/13-1-01-4_final_report.pdf.

Pinault, L., Tjepkema, M., Crouse, D. L., Weichenthal, S., van Donkelaar, A., Martin, R. V., et al. (2016). Risk estimates of mortality attributed to low concentrations of ambient fine particulate matter in the Canadian community health survey cohort. *Environmental Health: A Global Access Science Source*, *15*(1), 18. https://doi.org/10.1186/s12940-016-0111-6.

Pitchford, M. L., & Malm, W. C. (1994). Development and applications of a standard visual index. *Atmospheric Environment*, *28*(5), 1049–1054. https://doi.org/10.1016/1352-2310(94)90264-X.

Pitchford, M., Malm, W., Schichtel, B., Kumar, N., Lowenthal, D., & Hand, J. (2007). Revised algorithm for estimating light extinction from IMPROVE particle speciation data. *Journal of the Air and Waste Management Association*, *57*, 1326–1336.

Pope, C. A. 3rd, & Dockery, D. W. (2006). Health effects of fine particulate air pollution: Lines that connect. *Journal of the Air & Waste Management Association (1995)*, *56*(6), 709–742. https://doi.org/10.1080/10473289.2006.10464485.

Pope, C. A. (2007). Mortality effects of longer term exposures to fine particulate air pollution: Review of recent epidemiological evidence. *Inhalation Toxicology*, *19*(sup1), 33–38. https://doi.org/10.1080/08958370701492961.

Pope, C. A., Cropper, M., Coggins, J., & Cohen, A. (2015). Health benefits of air pollution abatement policy: Role of the shape of the concentration–response function. *Journal of the Air & Waste Management Association*, *65*(5), 516–522. https://doi.org/10.1080/10962247.2014.993004.

Prestemon, J. P., Hawbaker, T. J., Bowden, M., Carpenter, J., Scranton, S., Brooks, M. T., Sutphen R, Abt KL. (2013). Wildfire ignitions: a review of the science and recommendations for empirical modeling (General Technical Report No. SRS-171). Asheville, NC: USDA Forest Service, Southern Research Station.

Prestemon, J. P., Shankar, U., Xiu, A., Talgo, K., Yang, D., Dixon, E., et al. (2016). Projecting wildfire area burned in the south-eastern United States, 2011–60. *International Journal of Wildland Fire*, *25*(7), 715–729. https://doi.org/10.1071/WF15124.

Punger, E. M., & West, J. J. (2013). The effect of grid resolution on estimates of the burden of ozone and fine particulate matter on premature mortality in the United States. *Air Quality, Atmosphere and Health*, *6*(3), 563–573. https://doi.org/10.1007/s11869-013-0197-8.

Randerson, J. T., Chen, Y., van der Werf, G. R., Rogers, B. M., & Morton, D. C. (2012). Global burned area and biomass burning emissions from small fires. *Journal of Geophysical Research*, *117*, G04012. https://doi.org/10.1029/2012JG002128.

Reid, C. E., Brauer, M., Johnston, F. H., Jerrett, M., Balmes, J. R., & Elliott, C. T. (2016). Critical review of health impacts of wildfire smoke exposure. *Environmental Health Perspectives*, *124*(9), 1334–1343. https://doi.org/10.1289/ehp.1409277.

Spracklen, D. V., Mickley, L. J., Logan, J. A., Hudman, R. C., Yevich, R., Flannigan, M. D., & Westerling, A. L. (2009). Impacts of climate change from 2000 to 2050 on wildfire activity and carbonaceous aerosol concentrations in the western United States. *Journal of Geophysical Research*, *114*, D20301. https://doi.org/10.1029/2008JD010966.

Stocker, T. F., Qin, D., Plattner, G.-K., Alexander, L. V., Allen, S. K., Bindoff, N. L., et al. (2013). Technical summary. In T. F. Stocker, D. Qin, G.-K. Plattner, M. Tignor, S. K. Allen, J. Boschung, et al. (Eds.), *Climate Change 2013: The Physcial Science Basis* (pp. 79–113). Contribution of Working Group I to the Fifth Assessment Report of the Ingergovernmental Panel on Climate Change. Cambridge, UK and New York: Cambridge University Press.

Streets, D. G., Bond, T. C., Lee, T., & Jang, C. (2004). On the future of carbonaceous aerosol emissions. *Journal of Geophysical Research*, *109*, D24212. https://doi.org/10.1029/2004JD004902.

Sun, J., Fu, J. S., Huang, K., & Gao, Y. (2015). Estimation of future PM2.5- and ozone-related mortality over the continental United States in a changing climate: An application of high-resolution dynamical downscaling technique. *Journal of the Air & Waste Management Association (1995)*, *65*(5), 611–623. https://doi.org/10.1080/10962247.2015.1033068.

Tilmes, S., Lamarque, J.-F., Emmons, L. K., Kinnison, D. E., Ma, P.-L., Liu, X., et al. (2015). Description and evaluation of tropospheric chemistry and aerosols in the Community Earth System Model (CESM1.2).

Geoscientific Model Development, *8*(5), 1395–1426. https://doi.org/10.5194/ gmd-8-1395-2015.

US EPA. (2003). Guidance for Estimating Natural Visibility Conditions Under the Regional Haze Rule (Technical Report No. EPA 454/B-03-005). Research Triangle Park, NC: US Environmental Protection Agency Office of Air Quality Planning and Standards Emissions, Monitoring and Analysis Division Aid Quality Trends Analysis Group.

US EPA. (2006). Regulatory Impact Analysis for 2006 National Ambient Air Quality Standards for Particle Pollution. Technology Transfer Network Economics and Cost Analysis Support, US EPA. Retrieved from https://www.regulations.gov/document?D=EPA-HQ-OAR-2006-0834-0048.

US EPA. (2012). Our Nation's Air: Status and Trends Through 2010 (No. EPA-454/R-12-001). Research Triangle Park, NC: US Environmental Protection Agency Office of Air Quality Planning and Standards. Retrieved from https://www.epa.gov/sites/production/files/2017-11/documents/trends_brochure_2010.pdf.

Val Martin, M., Heald, C. L., Lamarque, J.-F., Tilmes, S., Emmons, L. K., & Schichtel, B. A. (2015). How emissions, climate, and land use change will impact mid-century air quality over the United States: A focus on effects at national parks. *Atmospheric Chemistry and Physics*, *15*(5), 2805–2823. https://doi.org/10.5194/acp-15-2805-2015.

Val Martin, M., Pierce, J. R., & Heald, C. L. (2018). Global fire emissions, fire area burned and air quality data projected using a global Earth system model (RCP45/SSP1 and RCP8.5/SSP3). Fort Collins, CO: Forest Service Research Data Archive. Updated 15 June 2018. doi:https:// doi.org/10.2737/RDS-2018-0021.

van Vuuren, D. P., Edmonds, J., Kainuma, M., Riahi, K., Thomson, A., Hibbard, K., et al. (2011). The Representative Concentration Pathways: An overview. *Climatic Change*, *109*(1–2), 5–31. https://doi.org/10.1007/s10584-011-0148-z.

Westerling, A. L. (2016). Increasing western US forest wildfire activity: Sensitivity to changes in the timing of spring. *Philosophical*

Transactions of the Royal Society B, *371*(1696), 20150178. https://doi.org/10.1098/rstb.2015.0178.

Westerling, A. L., & Bryant, B. P. (2008). Climate change and wildfire in California. *Climatic Change*, *87*(S1), 231–249. https://doi.org/10.1007/s10584-007-9363-z.

Westerling, A. L., Hidalgo, H. G., Cayan, D. R., & Swetnam, T. W. (2006). Warming and earlier spring increase western U.S. forest wildfire activity. *Science (New York, N.Y.)*, *313*(5789), 940–943. https://doi.org/10.1126/science.1128834.

Wiedinmyer, C., Akagi, S. K., Yokelson, R. J., Emmons, L. K., Al-Saadi, J. A., Orlando, J. J., & Soja, A. J. (2011). The Fire INventory from NCAR (FINN): A high resolution global model to estimate the emissions from open burning. *Geoscientific Model Development*, *4*(3), 625–641. https://doi. org/10.5194/gmd-4-625-2011.

Yue, X., Mickley, L. J., & Logan, J. A. (2014). Projection of wildfire activity in southern California in the mid-twenty-first century. *Climate Dynamics*, *43*(7–8), 1973–1991. https://doi.org/10.1007/s00382-013-2022-3.

Yue, X., Mickley, L. J., Logan, J. A., & Kaplan, J. O. (2013). Ensemble projections of wildfire activity and carbonaceous aerosol concentrations over the western United States in the mid-21st century. *Atmospheric Environment*, *77*, 767–780. https://doi.org/10.1016/j.atmosenv.2013.06.003.

The Washington Post
The Post's View Opinion

WE WON'T STOP CALIFORNIA'S WILDFIRES IF WE DON'T TALK ABOUT CLIMATE CHANGE

By Editorial Board
August 8

CALIFORNIA, THE nation's most populous state and the vvorld's fifth-largest economy, is on fire. In a state already known for monster conflagrations, the past month has been unusually destructive. The Mendocino Complex fire north of San Francisco is now officially the largest in California's history, having burned an area about the size of Los Angeles, and it is just one of the major blazes the state has had to face since last October.

President Trump tried to lay the blame on "bad environmental laws" and wasted water, claims that expe1ts quickly debunked. The 14,000 firefighters on the ground do not lack for water; they are battling blazes next to big lakes and other major bodies of water. The state's big rivers have not been "diverted into the Pacific," as Mr. Trump claimed; they flow into the ocean as they always have, though with large amounts sent to cities and farmland for human use.

Should even more of that water be taken to keep wild plants and soil moist, and therefore more resistant to fire? That wouldn't work. "Even if you built a massive statewide sprinkler system and drained all of our natural water bodies to operate it, it wouldn't keep up '¼1th evaporation from warmer temperatures from climate change," University of California at Merced professor LeRoy Westerling explained to NPR.

As much as the president might prefer to point fingers elsewhere, it is impossible to talk about California's blazes without considering the role of climate change. Four of the five largest conflagrations the state has had to ba le have come since 2012, according to the Los Angeles Times, and that is probably no mere coincidence. Droughts, storms and heat waves have occurred throughout history, of course, and it is hard to attribute any single event to climate change. But scientists have concluded that climate change has increased the frequency of extreme weather and will continue to do so.

In California, a half-decade-long drought was followed by swamping 'ISinter rains in 2016 and 2017, which encouraged rapid plant growth. Then, intense heat last summer dried out the land. That resulted in massive fires.last October. Come July, triple-digit heat once again fueled huge blazes, as arid land served as an ideal tinderbox. The state may offer an alarming taste of the troubles to come.

Even after major periods of rain, uninterrupted high heat can produce arid conditions quicldy1 and arid conditions lead to big fires. A 2016 study published in the Proceedings of the National Academy of Sciences concluded that human-caused climate change is responsible for about half the additional drying that researchers have found since the 1970s, resulting in a doubling of the area forest fires have consumed since 1984. Climate change may also increase lightning strikes, which are a major source of v-rildfires, and generate the high winds that can drive big blazes.

Meanwhile, earlier springtime melting means that the land has more time to dry out over the warmer months. Global warming will increasingly prime the environment for spectacular disasters.

Addressing global warming and hiring more firefighters are obvious responses; the federal government should also prepare to spend more money in disaster relief. Yet, pumping cash into ever-more firefighting is in part howforest fires got so bad in the West. So much of the U.S. Forest Service's budget has gone to firefighting that too little has been left for care and restoration. Lawmakers should examine the many ways they can help prevent another summer like this one - or worse.

Read More

Peter Gleick: Trump's nonsense tweets on water and wildfires are dangerous. David Arkush and David Michaels: Climate change isn't just cooking the planet. It's cooking our workforce.

The Post's View: There's still hope on global warming - if the world gets to work.

Robert J. Samuelson: Trump ignores the messy reality of global warming - and makes it all about him.

The Post's View: The practically cost-free way to slow global warming that Trun1p won't adopt.

The Washington Post
The story must be told.
Your subscription supports journalism that matters.

The New York Times

Trump Inaccurately Claims California Is Wasting Water as Fires Burn

By Lisa Friedman
Aug. 6, 2018

In his first remarks on the vast California wildfires that have killed at least seven people and forced thousands to flee, President Trump blamed the blazes on the state's environmental policies and inaccurately claimed that water that could be used to fight the fires was "foolishly being diverted into the Pacific Ocean."

State officials and firefighting experts dismissed the president's comments, which he posted on Twitter. "We have plenty of water to fight these wildfires, but let's be clear: It's our changing climate that is leading to more severe and destructive fires," said Daniel Berlant, assistant deputy director of Cal Fire, the state's fire agency.

He and others said that Mr. Trump appeared to be referring to a perennial and unrelated water dispute in California between farmers and environmentalists. Farmers have long argued for more water to be allocated to irrigating crops, while environmentalists counter that the state's rivers would suffer and fish stocks would die.

[For the latest updates on the Mendocino Complex Fire, read this story.]

The president first addressed the fires late Sunday, writing on Twitter, "California wildfires are being magnified & made so much worse by the bad environmental laws which aren't allowing massive amount of readily available water to be properly utilized." He also referred to a debate in forest management about the effectiveness of removing trees and vegetation as a fire control method.

On Monday, Mr. Trump expanded on his comments in another tweet, for a second time referring to water being diverted into the ocean.

The remarks came hours after the White House declared the wildfires a "major disaster" and ordered that federal funding be made available to help recovery efforts.

You have 3 free articles remaining.
Subscribe to the Times

Is There a Water Shortage?

California does not lack water to fight the Carr Fire and others burning across the state, officials said.

Mr. Berlant of Cal Fire declined to speculate on the meaning of Mr. Trump's statement that water was not being "properly utilized."

Asked about that line and the president's claim that water was being diverted into the Pacific, a spokesman for Gov. Jerry Brown, Evan Westrup, said in an email, "Your guess is as good as mine."

The White House did not respond to requests for clarification on Mr. Trump's statement.

William Stewart, a forestry specialist at the University of California, Berkeley, said he believed Mr. Trump was referring to the battle over allocating water to irrigation versus providing river habitat for fish.

That debate has no bearing on the availability of water for firefighting. Helicopters lower buckets into lakes and ponds to collect water that is then used to douse wildfires, and there is no shortage of water to do so, Cal Fire officials said.

California water regulators are preparing to negotiate how much water from the Sacramento-San Joaquin River Delta should flow to California's farms and how much should flow down the river and to the ocean to ensure fish have enough fresh water to spawn and hatch. The issue has long pitted environmentalists against the state's farming communities.

During the 2016 presidential campaign, Mr. Trump took on the farmers' grievances in language similar to his tweets this week.

"You have a water problem that is so insane, it is so ridiculous, where they're taking the water and shoving it out to sea," he said during a May 2016 campaign rally in Fresno. "They have farms up here, and they don't get water."

Recently, California Republicans encouraged the Trump administration to weigh in on the issue, inviting Interior Secretary Ryan Zinke to the Central Valley to discuss water rights in the state's agricultural heartland.

"It's a pretty big story, but it's got nothing to do with the fires," Mr. Stewart said.

Does Removing Trees Control fires?

Mr. Trump raised another issue when he wrote that officials "must also tree clear to stop fire spreading." Scientists and forest experts said the president was referring to a valid and continuing debate.

The timber industry has argued that "thinning" forests — removing certain trees to improve the health of the remaining ones and diminish the plants and underbrush that fuel fires — reduces the risk of wildfires. Republicans in Congress have sought to loosen environmental restrictions to allow more thinning. Democrats and environmentalists argue the practice will open the door to expanded commercial logging and threaten wildlife.

California already has policies in place to address wildfire risk.

LeRoy Westerling, a management professor at the University of California, Merced, who studies wildfires, said that Mr. Trump's statement about fire-control efforts hit on an important issue, but that he wrongly placed the blame on California.

Professor Westerling noted that while federal funding for lowering wildfire risk has been tied up in budget negotiations, California has allocated $256 million this year.

That money is coming from a source the Trump administration finds troublesome: revenues from California's program to reduce planet-warming greenhouse gases. Under its market-based approach for curbing carbon emissions, California sets a ceiling for the total amount of carbon that can be emitted. Companies are then required to obtain permits to release carbon into the atmosphere.

The Trump administration opposes federal efforts to address climate change.

California is "spending millions and millions of dollars on this while the federal government is sitting on its hands," Professor Westerling said. "And all that money is being raised because we're putting a price on carbon."

What About Climate Change?

Scientists noted that Mr. Trump's statement didn't address the role climate change has played in creating a hotter and drier fire season. The president in the past has dismissed climate change as a hoax and his top cabinet officials have questioned the established science that rising global temperatures are caused by human activity.

Michael F. Wehner, a senior staff scientist at Lawrence Berkeley National Laboratory, said it was not possible to quantify precisely the likelihood that climate change is having an impact on forest fires, as can now be done with other extreme-weather events such as heat waves.

And, he said, it's not easy to weigh how much of the problem can be laid at the feet of forest-management practices. However, climate change is making summers longer and drier, which expands the wildfire season.

"To dismiss the role of climate change on these fires is simply incorrect," he said.

California fire officials on Monday said the Carr Fire in Shasta County had ravaged more than 160,000 acres while the Mendocino Complex fires grew overnight and had charred more than 273,000 acres across Mendocino, Lake and Colusa counties.

The White House's disaster declaration ordered federal funding be made available to help recovery efforts. "Assistance can include grants for temporary housing and home repairs, low-cost loans to cover uninsured property loses and other programs to help individuals and business owners recover from the effects of the disaster," a White House statement said.

A version of this article appears in print on Aug. 7, 2018, on Page A13 of the New York edition with the headline: Trump Inaccurately Claims California Is Diverting Water From Fires.

Fueled by Climate Change, Wildfires Erode Air Quality Gains

Such fires are causing spikes in fine particles that threaten human health

By Scott Waldman, E&E News on July 17, 2018

Ventura, California during Thomas Fire in December, 2017.
Credit: Al Seib Getty Images.

Advertisement

Fourteen years ago, University of Washington researcher Daniel Jaffe installed an air pollution monitor on a mountainside outside Eugene, Ore.

His intention was to measure pollution levels, with a particular focus on tracking emissions from China that drift into the United States in the spring. But in recent years, the monitor has unexpectedly produced a second and more urgent data set: tracking fine particle pollution from wildfires in the western United States.

"We spend more of our time not worrying about what's coming across the ocean but worrying about what's coming here," he said.

Climate change is not just increasing the likelihood of wildfires in some areas of the country; it's also erasing decades of air pollution gains in those same regions, according to a study published yesterday in the *Proceedings of the National Academy of Sciences*. It shows that wildfires are causing a spike in air pollution across the West.

Air Quality Impacts of Wildfires

A man walks along Ventura Ave. in Ventura, California during the Thomas Fire in December, 2017.
Credit: Wally Skalij *Getty Images.*

Global warming creates conditions that feed wildfires. It has led to earlier snowmelts in the West, increased temperatures in summer and spring and drier conditions, research shows. That has sparked more frequent wildfires that last longer. And that increase in wildfires has increased fine particle air pollution, according to the study.

In the United States, fine particulate matter (PM2.5) is generated by coal-burning power plants, automobiles and the manufacturing sector, among other sources. PM2.5 refers to fine particulates that are no more than 2.5 microns in diameter, or one-thirtieth the width of a human hair.

EPA's National Ambient Air Quality Standards have been successful at lowering human-caused PM2.5 standards across the country for years, researchers found. A 2017 study found that reducing fine particle pollution even slightly could save 12,000 lives annually, particularly among vulnerable populations such as the elderly and people with asthma.

"These findings suggest that lowering the annual NAAQS may produce important public health benefits overall, especially among self-identified racial minorities and people with low income," that study said.

The new research shows that natural factors, exacerbated by human-caused climate change, may have more of a role in declining air quality than previously realized.

Wildfires in Montana last year caused a spike in PM2.5, Jaffe said. Monitors picked up the highest levels of air pollution ever recorded in the United States during August and September in Montana. And climate change is likely to drive similar events in the future, mostly in the summer and in Western states plagued by more frequent fires, he said.

In some areas of the United States, fine particles from wildfires are driving a second round of pollution events. Weather patterns in the winter can see cold air mass entrapped under warm air, which snares fine particles and raises air pollution levels. Both Utah and Fairbanks, Alaska, typically record high levels of air pollution in the winter, but wildfires are driving a second spike in PM2.5 levels in the summer.

PM2.5 is detrimental to human health. It affects the lungs as well as the heart, according to EPA. It can aggravate asthma, decrease lung function, and lead to premature death in people with lung and heart disease.

Clean Air Act regulations have targeted PM2.5. Controls on coal-burning power plants and other sources of pollution have yielded significant reductions of pollution for years. Nationally, the amount of fine particle pollution has dropped 42 percent since 2016. However, researchers found that emissions have increased in parts of the country where wildfires are concentrated.

Scientists examined data from rural monitoring sites. They found increases in PM2.5 in all or parts of Montana, Idaho, Oregon, California, Nevada, Utah and Wyoming.

They found an increase for total carbon, which indicates wildfire emissions, in the Northwest. By contrast, the rest of the country saw decreasing total carbon, researchers found.

"If you've been out in the West at all in the last few years, we've just seen more and more fires and bigger fires, so that was why we went looking for it," Jaffe said. "I guess what surprised us was the geographic extent over which we could really pull out a statistically significant increase."

The results show that fine particle pollution from wildfires may be worse than researchers have realized, Jaffe said. What's more, they come as EPA has proposed a new science transparency rule that would restrict the use of key air pollution studies in crafting regulations. Critics have noted that the proposed transparency rule would exclude future consideration of a landmark 1993 Harvard University air pollution study—the "Six Cities" paper—that has been used to back air regulations for decades.

Jaffe said snowpack has a direct effect on the wildfire season, especially if the snow extended into late June and early July and kept the soil wet. In addition, insects such as the pine bark beetle have damaged millions of trees, providing fuel for bigger fires that last longer, he said. Still, he cautioned that climate change is a threat multiplier for wildfire season but that forest management issues have also played a significant role in driving wildfires.

"We want to be careful not to put it all on climate change, but climate change is clearly a contributing factor, and particularly in the size of these fires," he said. "A fire that used to become a small fire has now become a massive conflagration."

With fine particle pollution from wildfires increasing, it may need even more consideration by future regulators, said Gannet Hallar, an atmospheric scientist at the University of Utah who was not involved in the study. She said it shows that the highest-polluting wildfire events are intensifying.

"The really important point of this paper is that these events, although episodic, are increasing in their intensity," she said.

Reprinted from Climatewire with permission from E&E News. E&E provides daily coverage of essential energy and environmental news at www.eenews.net.

About the Author(S)

Scott Waldman

Recent Articles

As Climate Scientists Speak Out, Sexist Attacks Are on the Rise Coming Soon: Acting EPA Administrator's First Big Moves on Science Car Rules Fight Pits Safety against Pollution

E&E News

Recent Articles

North Carolina's Natural Hurricane Defenses Are Disappearing Want a Carbon Tax? Wait Until Next Year, Advocates Say

This Federal Lab Works to Make Cars More Efficient, as Trump Pumps the Brakes

Latest News

Policy & Ethics

Don't Condemn People Who Don't Evacuate for Hurricane Florence
4 hours ago — Mika McKinnon

Space

Space Station Commander: It's "Absolutely a Shame" to Suggest Astronauts Caused Leak
4 hours ago — Mike Wall and SPACE.com

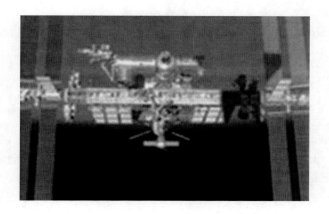

Weather

North Carolina's Natural Hurricane Defenses Are Disappearing
4 hours ago — Chelsea Harvey and E&E News

Behavior and Society

7 Ways to Let Go of Guilt
5 hours ago — Savvy Psychologist Ellen Hendriksen

Natural Disasters
Health Officials Rush to Protect Seniors, the Most Vulnerable Group, from Hurricane Florence

6 hours ago — Liz Szabo, JoNel Aleccia, Doug Pardue and STAT

Medical and Biotech
Study Cracks Open the Secrets of Genetic Mutations That Boost Breast and Ovarian Cancer Risk

6 hours ago — Sharon Begley and STAT

NEWSLETTER

Expertise. Insights. Illumination.

SUBSCRIBE NOW!
FOLLOW US
Store
About
Press Room
More

Scientific American is part of Springer Nature, which owns or has commercial relations with thousands of scientific publications (many of them can be found at www.springernature.com/us). Scientific American maintains a strict policy of editorial independence in reporting developments in science to our readers.

© 2018 SCIENTIFIC AMERICAN, A DIVISION OF SPRINGER NATURE AMERICA, INC.
ALL RIGHTS RESERVED.

MEGAFIRES: THE GROWING RISK TO AMERICA'S FORESTS, COMMUNITIES, AND WILDLIFE

Copyright © 2017 National Wildlife Federation

Lead Authors: Shannon Heyck-Williams, Lauren Anderson, and Bruce A. Stein.

We appreciate the work and dedication to conservation of all National Wildlife Federation staff and our affiliate partners, who help make efforts like this possible. In particular, we would like to thank the following for their contributions to this report: Chris Adamo, Mitch Friedman, Patty Glick, Mike Leahy, Jim Lyon, Jim Murphy, Collin O'Mara, Josh Saks, and Tiffany Woods.

Suggested citation: Heyck-Williams, S., L. Anderson, and B.A. Stein. 2017. *Megafires: The Growing Risk to America's Forests, Communities, and Wildlife.* Washington, DC: National Wildlife Federation.

Cover image: Blackhall fire, Wyoming, 2000. Photo: Kerri Greer

The National Wildlife Federation is the education partner for *Years of Living Dangerously,* an Emmywinning climate change film series. You can learn more about wildfires and climate change science with our lesson plans, videos, and resources on www.climateclassroom.org.

Megafires is available online at: www.nwf.org/megafires-report

Elk taking refuge in river, Sula fire, Montana, 2000. Photo: John McColgan, USFS.

President's Letter

As I write this, explosive wildfires are raging across nearly 200,000 acres in northern California, devastating communities and pushing the total area burned in 2017 to more than 8.5 million acres nationally. Although wildfires are a natural and essential feature of many forest ecosystems, there has been a dramatic increase in recent decades of unusually large and severe "megafires." There are multiple reasons for this increase in extreme wildfires. Changing climatic conditions—including earlier onset of spring, earlier snow melt, hotter summer temperatures, and prolonged drought—are reducing moisture levels in soil and forest vegetation. The spread of pest species like bark beetle, often facilitated by a changing climate, is weakening the resilience of many natural ecosystems. And seriously overgrown forests, resulting from years of fire suppression and coupled with insufficient resources and bureaucratic obstacles for proactive forest management and restoration, have created conditions that are ripe for explosive megafires.

Fighting wildfires is now devouring more than half of the U.S. Forest Service budget, depriving the agency of critical resources for restoration and improved forest management that would reduce ongoing fire risks. In recent years the agency has been spending more than $1 billion annually to fight wildfires,[27] and in 2015, one of the worst years for wildfires on record, the U.S. government as a whole spent more than $2.6 billion[28]—a record that we are poised to match this year. Despite increasing fire activity and escalating costs, Forest Service wildfire budgets are based on historical averages rather than future projections. As a result, as fire season progresses and budgeted wildfire funds are exhausted, money is shifted ("borrowed") from other activities, including recreation, wildlife management, and forest restoration. But in contrast to hurricanes, tornadoes, and major floods, disaster funds are not available to cover the exceptional costs involved in fighting these catastrophic megafires.

The social, economic, and ecological costs of these megafires has been devastating. Communities have lost thousands of buildings, suffered tens of

---

[27] National Interagency Fire Center. 2017. Federal Firefighting Costs. Accessed October 4, 2017. https://www.nifc.gov/fireInfo/fireInfo_documents/SuppCosts.pdf.
[28] Streater, S. 2016. Policymakers must prepare for larger blazes – report. Greenwire, January 18, 2016. https://www.eenews.net/greenwire/stories/1060030747.

billions of dollars in damages, and, tragically, people have died. Waterways and other habitats have been degraded, imperiling fish and wildlife. Tens of millions of tons of climate-altering carbon dioxide have been released. Unfortunately, current wildfire policy is woefully insufficient to address this urgent crisis.

Our nation needs bipartisan Congressional leadership to fix the forest fire funding crisis and ensure adequate funding for restoring forests and appropriately fighting wildfires. We must accelerate the pace of forest restoration by promoting outcome-driven, collaborative processes, by expanding the use of prescribed fire, and by improving environmental review of beneficial restoration projects. We must boost the resilience of our forests and communities, discourage development in fire-prone areas, and prioritize fire risk reduction in the wildland-urban interface. And, Congress must confront the underlying causes of climate change that are exacerbating the wildfire crisis by ensuring sufficient funding for climate research and reducing climate-altering pollution.

The time for bipartisan action is now.

Collin O'Mara
President and CEO
National Wildlife Federation

MEGAFIRES: A GROWING THREAT

Druid fire, Yellowstone National Park, 2013. Photo: Mike Lewelling, NPS.

Wildfire, like actors in a Greek drama, has two opposing faces: one of destruction and another of rejuvenation. Scenes of smoldering forests and burnt-out homes vividly illustrate the destructive force of uncontrolled wildfires. Paradoxically, many U.S. forests are not only adapted to burn periodically, but actually depend on fire for their rejuvenation, maintenance, and health.

Consequently, forest and wildfire management is and always has been an extraordinarily complex and high stakes issue, which rapid climate change is making even more challenging. In particular, the increase in large, intensely hot "megafires" not only poses increased risks to local communities and economies, but has the ability to permanently transform the ecosystems and habitats through which they burn, with profound implications for wildlife.

The current crisis in wildfire and forest management has its roots in three interacting dynamics: the legacy of past forest management and fire suppression; dramatic increases in housing development in the fire-prone wildland-urban interface; and rapidly changing climatic conditions. Reducing risks from megafires will require that we address each of these underlying problems, including: scaling up efforts to tackle the massive backlog in forest restoration; encouraging more responsible and fire-wise development in wildland areas; and confronting climate change both by reducing greenhouse gases and by incorporating climate considerations in forest management and restoration.

What is most urgently needed, however, is to fix the broken federal budget process for fighting wildfires.

> *The increase in large, intensely* hot "megafires" not only poses *increased risks to local communities* and economies, but has the ability to permanently transform the ecosystems and habitats through which they burn, with profound implications for wildlife.

More, Bigger, and Hotter Fires

Eagle Creek fire along Columbia River Gorge, 2017. Photo: Christian Roberts-Olsen, Shutterstock.

For several years California and other parts of the West have been in a severe and sustained drought, and the La Niña weather of last winter was anticipated to reduce the risk of severe wildfire in the region by bringing more rain and snow. Intense summer heatwaves, however, countered many of the expected benefits, and the explosive growth of annual grasses stimulated by winter rains may have even added to the fire risk. These conditions set the stage for the deadly wildfires now scorching nearly 200,000 acres across California and devastating many local communities. Indeed, 2017 is on track to be one of the most active years for U.S. wildfires on record, with a total of 8.5 million acres burned, and more than 21,000 firefighters assigned to wildfires in 10 western states.[29]

Over 1.2 million acres have burned in Montana alone.[30] The Lolo Peak fire has burned over 50,000 acres, covering much of the surrounding areas in Montana and Idaho with smoke. The Lodgepole Complex and Rice Ridge fires have also scorched hundreds of thousands of acres in Montana. In

---

[29] National Interagency Fire Center. 2017. National Fire News, October 12, 2017. https://www.nifc.gov/fireInfo/nfn.htm.

[30] Northern Rockies Coordination Center. 2017. Year-to-Date Fires and Acres by State and Agency. Accessed October 4, 2017. https://gacc.nifc.gov/nrcc/predictive/intelligence/ytd_historical/ytd-daily-state.htm.

California, a state of emergency was declared for Los Angeles County as the La Tuna fire blanketed an area near Burbank, requiring over 1,000 firefighters to protect the heavily populated area.[31] In Oregon, the Eagle Creek wildfire has burned nearly 50,000 acres across the iconic Columbia River Gorge.[32] In Washington State, the Diamond Creek fire has charred over 125,000 acres.[33]

These fires not only affected large areas of wildlife habitat, but have had direct implications for wildlife managers and their efforts to restore key wildlife populations. In Washington State, for instance, biologists raced to rescue a population of endangered pygmy rabbits as the Sutherland Canyon fire threatened to overtake their captive breeding enclosure.[34] Thick smoke and the threat of new fires in Montana forced federal wildlife managers to postpone the annual roundup at the National Bison Range.[35] Western wildfires in 2017 have also had serious impacts on people and communities, blanketing cities in ash, and in many areas making the air dangerous to breathe.

With massive fires continuing to cover the West, experts are observing some disturbing trends. There is strong evidence that regional warming and drying in the western United States is linked to increased fire frequency and size, as well as to longer fire seasons.[36] The U.S. Forest Service, for example, has concluded that fire seasons are now on average 78 days longer than in

---

[31] Chavez, N., J. Sutton and D. Andone. 2017. More than 1,000 firefighters battling largest fire in Los Angeles history. CNN, September 4, 2017. http://www.cnn.com/2017/09/02/us/los-angeles-wildfire/index.html.
[32] InciWeb: Incident Information System. 2017. Accessed October 4, 2017. https://inciweb.nwcg.gov/.
[33] Ibid.
[34] Camden, J. 2017. Rescuing endangered rabbits from fire danger. Spokane-Review, September 13, 2017. http://www.spokesman.com/stories/2017/sep/13/rescuing-endangered-rabbits-from-fire-danger/#/0.
[35] Backus, P. 2017. Fire and smoke postpone annual Bison Range roundup. Missoulian, Sep 13, 2017. http://missoulian.com/news/local/fire-and-smoke-postpone-annual-bison-range-roundup/article_6ffc5d05-1d63-5e32-93dd-ffb126b814bb.html.
[36] Balch, J.K., B.A. Bradley, J.T. Abatzoglou, R.C. Nagy, E.J. Fusco, and A.L. Mahood. 2017. Human-started wildfires expand the fire niche across the United States. Proceedings of the National Academy of Sciences 114: 2946-2951.

1970.[37] The area of forest burned annually in the Pacific Northwest has increased by nearly 5,000 percent since the early 1970s, while the area burned in the Southwest has increased by nearly 1,200 percent.[38,39]

It is not only the frequency and size of fires that are changing, but also their intensity and severity. Indeed, more, bigger, and hotter wildfires are becoming the new normal. Extreme fire behavior—characterized by rapid fire spread, intense burning, prolific crowning, strong convection columns, and unpredictable shifts—not only can have serious ecological effects, but increasingly puts wildland firefighters at risk.

Decades of fire suppression has in many places prevented smaller, less-intense surface fires that help to naturally thin forests. As a result, many forests have grown so dense that once ignited, flames quickly climb understory "ladder fuels" and set the tree canopies ablaze. Crown fires can burn so hot they have the ability to create their own weather, spreading the fire ever further and hindering control efforts. Complicating matters more, there has been an increase in the number of human-caused wildfire ignitions, due in part to the dramatic expansion of housing into the wildland-urban interface. Nearly 45 million housing units are now in naturally fire-prone wildland areas,[40] and one study estimates that human-caused ignition is responsible for 84 percent of all U.S. wildfires.[41]

Rather than leaving a tapestry of burned and unburned areas, which facilitates forest regeneration and provides a diversity of habitats for wildlife, ultra-hot megafires can destroy complete forest stands and burn across entire landscapes. If hot enough, extreme fires can even sterilize the soil by killing subsurface seed banks that normally aid in post-fire recovery.

---

[37] U.S. Forest Service. 2015. *The Rising Cost of Wildfire Operations: Effects on the Forest Service's Non-Fire Work*. Washington, DC: U.S. Forest Service, Department of Agriculture. https://www.fs.fed.us/sites/default/files/2015-Fire-Budget-Report.pdf.

[38] Westerling, A.L. 2017. Wildfires in West have gotten bigger, more frequent and longer since the 1980. The Conversation, May 23, 2016. https://theconversation.com/wildfires-in-west-have-gotten-bigger-more-frequent-and-longer-since-the-1980s-42993.

[39] Westerling A.L., H.G. Hidalgo D.R. Cayan and T.W. Swetnam. 2006. Warming and earlier spring increase western U.S. forest wildfire activity. Science 313: 940-943.

[40] Radeloff V.C., R.B. Hammer, S.I. Stewart, J.S. Fried, S.S. Holcomb and J.F. McKeefry. 2005. The wildland–urban interface in the United States. Ecological Applications 15: 799-805.

[41] Balch, J.K., B.A. Bradley, J.T. Abatzoglou, R.C. Nagy, E.J. Fusco and A.L. Mahood. 2017. Human-started wildfires expand the fire niche across the United States. Proceedings of the National Academy of Sciences 114: 2946-2951.

Indeed, forests in some places may never recover from these fires and instead be permanently transformed into shrubland or grasslands.[42] For example, in 2011 the Las Conchas fire in New Mexico burned more than 156,000 acres of forest and scrubland, one of the largest fires in the state's history. The fire burned so intensely that only bare dirt and tree stumps were left in many places, and some burned areas will probably never revert to forest.[43,44]

An increase in the frequency of wildfire events can also lead to the long-term conversion of forests and other wildlife habitats by enabling the expansion of non-native invasive species.[45] Many invasive plants are able to rapidly colonize disturbed areas, often outcompeting native species. For example, in northern Nevada and elsewhere in the Great Basin, burned sagebrush ecosystems, on which sage grouse and migrating mule deer depend, often convert to grasslands dominated by invasive cheatgrass, which has little wildlife value. Unfortunately, cheatgrass also provides the type of fine fuel that promotes even more frequent wildfires in these areas, leading to permanent conversion from native sagebrush steppe habitat to invasive grassland.[46] This conversion is not only a major contributor to the decline of sage grouse, but also effects the many other wildlife that depend on this habitat type, including short-horned lizard, sharptailed grouse, pygmy rabbit, and Brewer's sparrow.

Some wildlife may be able to adapt or even thrive in fire-altered or transformed habitats, or migrate to areas better supporting their needs. Others, though, will be negatively impacted either as an immediate effect of

---

[42] Millar, C.I. and N.L. Stephenson. 2015. Temperate forest health in an era of emerging megadisturbance. Science 349: 823-826.

[43] Nijhuis, M. 2012. Forest fires: burnout. Nature 489: 352-354.

[44] Haussamen, H. 2014. New Mexico's forests are warming and transforming. New Mexico In Depth, August 30, 2014. http://nmindepth.com/2014/08/30/new-mexicos-forests-are-warming-and-transforming/.

[45] Keeley, J.E., M. Baer-Keeley and C.J. Fotheringham. 2005. Alien plant dynamics following fire in Mediterranean-climate California shrublands. Ecological Applications 15: 2109-2125.

[46] Balch, J.K., B.A. Bradley, C.M. D'Antonio and J. Gómez-Dans. 2013. Introduced annual grass increases regional fire activity across the arid western USA (1980–2009). Global Change Biology 19: 173–183.

a large fire, or in the aftermath when food resources, water, or shelter are hard to come by, or their habitats are permanently lost.[47]

Pygmy rabbit. Photo: USFWS.

How Climate Change Increases the Risk of Megafires

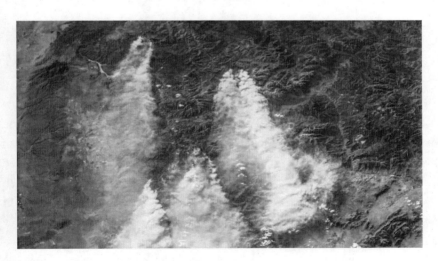

Wildfires in central Idaho, 2013. Photo: NASA.

---

[47] Zielinski, S. 2014. What do wild animals do in a wildfire? National Geographic, July 22, 2014. http://news.nationalgeographic.com/news/2014/07/140721-animals-wildlife-wildfires-nation-forests-science/.

Climate change increases the risk of more frequent and severe fire in several ways. Earlier spring snowmelt, higher temperatures in spring, summer, and fall, and increases in evapotranspiration, all contribute to drying of vegetation, and extend the geographic area and time periods in which forests become combustible.[48] These changing climatic conditions have resulted in wildfire seasons becoming longer, particularly in the western United States. In parts of California, fire season is now 50 days longer than in 1979,[49] while Alaska moved its official fire season, defined as the date permits are required for burning residential yard and other refuse, from May to April.[50] With fires burning both earlier and later, some places may even start experiencing what has been called "year-round fire season."

Higher temperatures and extreme drought can trigger tree stress and mortality, which can increase fire risk. In California, for instance, more than 100 million trees died during the recent prolonged drought. And as discussed below, drought-induced stress can exacerbate outbreaks of forest pests, such as bark beetles, which can also increase the susceptibility of forests to

---

[48] Westerling, A.L. 2017. Increasing western U.S. forest wildfire activity: sensitivity to changes in the timing of spring. Philosophical Transactions of the Royal Society B 371: 20150178.

[49] Jolly, W.M., M.A. Cochrane, P.H. Freeborn, Z.A. Holden, T.J. Brown, G.J. Williamson and D.M.J.S. Bowman. 2015. Climate-induced variations in global wildfire danger from 1979 to 2013. Nature Communications 6: 7537.

[50] Richtel, M. and F. Santos. 2016. Wildfires, once confined to a season, burn earlier and longer. New York Times, April 12, 2016. https://www.nytimes.com/2016/04/13/science/wildfires-season-global-warming.html?_r=0.

wildfire.[51] Drought conditions can also contribute to super-hot fires that produce more lasting damage, where high temperatures penetrate deeper into soils and prevent seeds from germinating once the fire is over.

Although there are multiple reasons for the overall increase in wildfire activity, researchers have concluded that over the past few decades climate change has caused more than half the increase in fuel aridity, and is responsible for a doubling in the cumulative forest area burned.[52] Looking to the future, researchers project that climate change will increase the potential for very large fires, both through increasingly frequent conditions conducive to these fires (i.e., changes in temperature, precipitation, and relative humidity) and through lengthening of the seasonal window when fuels and weather support these fires.[53,54]

---

[51] Joyce, L., et al. 2014. Forests. p. 175-194 in: J.M. Melillo, T.C. Richmond, and G.H. Yohe, eds. *Climate Change Impacts in the United States: The Third National Climate Assessment*, U.S. Global Change Research Program. Washington, DC: Government Printing Office.

[52] Abatzogloua, J.T. and A.P. Williams. 2016. Impact of anthropogenic climate change on wildfire across western US forests. Proceedings of the National Academy of Sciences: 113: 11770–11775.

[53] R. Barbero, R., J.T. Abatzoglou, N.K. Larkin, C.A. Kolden and B. Stocks. 2015. Climate change presents increased potential for very large fires in the contiguous United States. International Journal of Wildland Fire 24: 892-899.

[54] Stavros, E.N., J.T. Abatzoglou, D. McKenzie and N.K. Larkin. 2014. Regional projections of the likelihood of very large wildland fires under a changing climate in the contiguous Western United States. Climatic Change 126: 455-468.

King fire, California, 2014. Photo: USFS.

Not only does climate change increase the risk of extreme fires, but megafires in turn contribute to the underlying cause of climate change through the release of large quantities of carbon into the atmosphere. In Washington State, for instance, wildfires were the second largest source of carbon dioxide emissions in 2015, behind only the transportation sector.[55] Although wildfires always release carbon, the amount released through most low intensity fires typically is offset in subsequent years by vegetation regrowth and recovery in the burnt area. In contrast, the massive amounts of carbon released by megafires, coupled with declines in the capacity of the landscape to recover, may in some instances lead to a shift from carbon "sink" to carbon "source."[56]

---

[55] Siemann, D., R. Ottmar, J. Halofsky and D. Donato. 2016. *Washington's Greenhouse Gas Emissions: Comparison of 2012 Human Caused Emissions with 2015 Wildfire Emissions.* Washington State Department of Natural Resources. https://climatetrust.org/pacific-northwest-forests-carbon-sink-or-carbon-source/.

[56] Wear, D.N. and J.W. Coulston. 2015. From sink to source: regional variation in U.S. forest carbon futures. Scientific Reports 5: 16518.

> *Researchers have concluded that* over the past few decades climate change has caused more than half the increase in fuel aridity, and is responsible for a doubling in the cumulative forest area burned.

From Fires to Floods

Because severe fires can burn away much of the vegetation that holds soil in place and retains run-off, when the rainy season returns there is often an increased risk of flooding below the burned area. For example, the 2010 Schultz fire burned just over 15,000 forested acres above Flagstaff, Arizona and caused the evacuation of hundreds of homes. Monsoon rains following the fire caused heavy flooding that resulted in extensive property damage in Flagstaff and took the life of a young girl. Overall, estimates of the total impact of the Schultz fire place the cost between $133 million and $147 million. These costs are a heavy burden on rural communities, and to reduce the risk of future fire and flood disasters, the Flagstaff community passed a $10 million bond to finance wildfire control treatments throughout local watersheds, including on federal lands.[57]

Runoff from burned areas can also result in soil and ash polluting streams and rivers. The ash can increase the pH level in water, and sediment can clog the gills of fish as well as destroy and degrade fish habitat. Studies

---

[57] Small-Lorenz, S.L., B.A. Stein, K. Schrass, D.N. Holstein and A.V. Mehta. 2016. *Natural Defenses in Action: Harnessing Nature to Protect Our Communities*. Washington, DC: National Wildlife Federation. http://www.nwf.org/nature-in-action.

in Arizona show that the state fish, the Apache trout, can suffer severe population declines following major wildfires.⁵⁸

> *Monsoon rains following the* fire caused heavy flooding that resulted in extensive property *damage in Flagstaff and took the* life of a young girl.

Schultz fire above Flagstaff, Arizona. Photo: Brady Smith, USFS.

Aftermath of Schultz fire. Photo: Bill Morrow.

---

⁵⁸ Long, J. 2008. Persistence of Apache trout following wildfires in the White Mountains of Arizona. p. 219-234 in: C. van Riper III and M.K. Sogge eds., *The Colorado Plateau III: Integrating Research and Resources Management for Effective Conservation.* Tucson: University of Arizona Press. https://www.fs.fed.us/psw/publications/jwlong/psw_2008_long001.pdf.

Apache trout. Photo: USFWS.

Beetle Infestations Can Amplify Fire Risk

Mountain pine beetle infestation, Colorado. Photo: Hustvedt.

Fire is not the only major, climate-amplified disturbance affecting U.S. forests. Bark beetles are naturally occurring forest insects that have reached unprecedented epidemic levels over the past two decades. Mountain pine beetle infestations in particular have caused significant tree mortality across millions of acres in the Rocky Mountains. These beetle outbreaks have been correlated with climatic changes, and especially warmer winter temperatures that allow more beetles to successfully overwinter.[59]

---

[59] Bentz, B.J., J. Régnière, C.J Fettig, E.M. Hansen, J.L. Hayes, J.A. Hicke, R.G. Kelsey, J.F. Negrón and S.J. Seybold. 2010. Climate change and bark beetles of the western United States and Canada: direct and indirect effects. BioScience 60: 602-613.

Various species of bark beetles are also causing significant forest damage elsewhere in the U.S., including Alaska and the Southeast, and appear to be expanding their range due to warming conditions. Southern bark beetles, for instance, historically have been restricted to pitch pine forests of the South, constrained by cold winters further north. That has now changed, and since 2002 the beetles have damaged more than 30,000 acres of forest in New Jersey, and recently have been detected in forests as far north as Massachusetts.[60]

In addition to the effect of rising winter and summer temperatures on beetle reproduction, drought conditions create water stress in forest trees and can make them more susceptible to bark beetle infestations.

As large swaths of forest succumb to beetle infestations, beetle-killed trees can increase the risk of wildfires, particularly early in an outbreak when dead or dying needles are still on the trees.[61] Colorado, for example, has recently experienced the largest bark beetle outbreak in its recorded history, which left hundreds of thousands of trees dead and vulnerable to wildfire.[62] In 2016, the Beaver Creek fire burned over 38,000 acres and cost an estimated $30 million to contain.[63] The fire was made more difficult to manage as it burned through beetle infested timber and dead trees.

---

[60] Schlossberg, T. 2016. Warmer winter brings forest-threatening beetles north. New York Times, March 18, 2016. https://www.nytimes.com/2016/03/22/science/southern-pine-beetles-new-england-forests.html?_r=0.

[61] Hicke, J.A., M.C. Johnson, J.L. Hayes and H.K. Preisler. 2012. Effects of bark beetle-caused tree mortality on wildfire. Forest Ecology and Management 271: 81-90.

[62] Streater, S. 2017. In Colo., fire season is now a yearlong problem. Greenwire, March 24, 2017. https://www.eenews.net/greenwire/stories/1060052053/search?keyword=wildfire.

[63] Paul, J. 2016. Beaver Creek fire cost at roughly $30 million as massive blaze near Walden continues to smolder. Denver Post, November 21, 2016. http://www.denverpost.com/2016/11/21/how-much-does-beaver-creek-fire-cost/.

The combination of beetle kill and fire can have serious effects on native wildlife. As an example, the 2016 Hayden Pass fire in Colorado's Sangre de Cristo range burned through an area badly affected by bark beetles. This area contains streams that are the only known refuge for a unique and isolated population of cutthroat trout, which was only discovered by biologists in 1996. Fall monsoon rains after the fire washed significant amounts of debris, ash, and sediment into the trout's habitat. Surveys following the fire and rains did not locate any remaining trout in Hayden Creek, and this unique trout might have been extirpated but for emergency rescue efforts following the fire that brought 158 of the fish into captivity for breeding and reintroduction.[64]

More Haze Follows Megafires

Western high country and wilderness are increasingly seeing the impacts of haze pollution from megafires, obscuring the views westerners and visitors alike cherish. New research indicates that more frequent drought and wildfire are leading to increased haze in western states as the fires produce a combination of small dust, soot, ash, smoke particles, and other air pollutants.[65] Small particles are public health concerns as they can lodge deep within the lungs and cause respiratory and cardiac distress and illness, and even premature death.[66]

---

[64] Wickstrom, T. 2016. Colorado wildfires' effects on wildlife, recreation and habitat. Denver Post, July 19, 2016. http://www.denverpost.com/2016/07/19/wickstrom-colorado-wildfires-effects-on-wildlife-recreation-and-habitat/.

[65] Finley, B. 2017. Hazier days in the high country, western U.S. due to drought and forest fires, scientists find. Denver Post, January 8, 2017. http://www.denverpost.com/2017/01/08/drought-forest-fires-visibility-western-wilderness/.

[66] Lipsett, M., B. Materna, S.L. Stone, S. Therriault, R. Blaisdell and J. Cook. 2008. Wildfire Smoke: A Guide for Public Health Officials. California Department of Public Health, U.S. Environmental Protection Agency, California Office of Environmental Health Hazard Assessment, Missoula County Health Department, California Air Resources Board. https://www.frames.gov/files/6213/7175/2941/Guide_pub_hlth_2013.pdf.

Smoke drifting over Las Vegas from Mount Charleston fire, 2013. Photo: Tomás Del Coro.

It is not only remote and wild places out West being impacted. Urban areas are also being affected by these fires. In 2014, the Carpenter 1 fire burned the landmark peak of Mount Charleston 35 miles northwest of Las Vegas. Smoke from the fire triggered a health advisory from the Clark County Department of Air Quality that lasted days and impacted the entire Las Vegas area.[67]

---

[67] Black, J. 2013. Smoke from massive wildfire billows over Las Vegas. NBC News, July 9, 2013. http://www.nbcnews.com/news/us-news/smoke-massive-wildfire-billows-over-las-vegas-v19378606.

A Growing Threat in the Southeast

Fire damage, Gatlinburg, Tennessee, 2016. Photo: Michael Tapp.

Extreme fires are not restricted to the arid West: the Southeast is also experiencing an increasing number of large and intense blazes. Many southeastern forests, especially softwood pines, are fire-adapted and depend on relatively frequent, low intensity burns for their maintenance.

A changing climate and associated periodic drought conditions, however, are contributing to much larger, more intense wildfires that can affect not just drier pinelands, but burn typically moist hardwood forests as well.

Although the southeast is in general characterized by warm, humid climates, severe drought conditions, such as occurred in 2007 and 2016, are creating conditions ripe for major fires. Last year, some of the biggest fires in the Southeast were in Georgia and North Carolina,[68] but the fire that caught the most national attention occurred in Tennessee. This explosive fire started in Great Smoky Mountains National Park as a result of human activities, and quickly spread to the gateway towns of Gatlinburg and Pigeon Forge, Tennessee. The Gatlinburg fire burned 1,700 structures, caused 14,000 to evacuate, and resulted in the death of 14 people, the deadliest the state has ever experienced.[69]

Forecasting the long-term impacts of climate change in the Southeast is challenging because of variability in precipitation patterns across the region. It is clear, however, that summer and winter temperatures in the region are rising. This will increase evapotranspiration and make drought conditions more likely during periods of low precipitation, with consequent increase in wildfire risk.[70]

Prescribed burns are an important and widely used management tool in the region, and are essential for maintaining and restoring the biologically rich longleaf pine forests, as well as for rebuilding populations of popular game species like bobwhite quail. To be effective and safe, these controlled burns must be carried out under particular weather conditions. Projected climate shifts are likely to shrink the availability of those conditions and significantly constrain the capacity for wildlife and natural resource managers to employ this essential tool for forest restoration and management.

---

[68] Dorman, T. 2016. West Coast crews mobilize to battle Southeast fires. USA Today, November 11, 2016. https://www.usatoday.com/story/news/nation-now/2016/11/16/west-coast-crews-mobilize-battle-southeast-fires/93948412/.

[69] Zenteno, R., J. Hanna and M. Park. 2016. 14 confirmed dead in Tennessee wildfires. CNN, December 5, 2016. http://www.cnn.com/2016/12/05/us/tennessee-gatlinburg-wildfires/index.html.

[70] Thompson, A. 2016. What a warmer future means for southeastern wildfires. Climate Central, November 23, 2016. http://www.climatecentral.org/news/warmer-future-southeastern-wildfires-20912.

Northern bobwhite quail. Photo: Dick Daniels.

Re-Establishing Natural Fire Regimes

While fire is a natural process in most U.S. forests, there is wide variation in the natural fire regimes that characterize different forest types. Fire regime refers to a combination of factors, such as the frequency, intensity, size, pattern, season, and severity of burns.[71] As an example, many low-elevation ponderosa pine forests historically had a fairly frequent fire return interval (every 10-30 years), with low-severity surface fires that resulted in relatively open forest conditions.[72] In contrast, northern Rockies lodgepole pine forests historically had long fire return cycles (often greater than 100 years), with high-severity, stand-replacing crown fires (as occurred, for example, in the 1988 Yellowstone fires). Understanding

---

[71] Brown, J.K. and J.K. Smith, eds. 2000. Wildland fire in ecosystems: effects of fire on flora. General Technical Report RMRS-GTR-42-vol. 2. Ogden, UT: U.S. Department of Agriculture, Forest Service, Rocky Mountain Research Station.
[72] Rocca, M.E., P.M. Brown, L.H. MacDonald and C.M. Carrico. 2014. Climate change impacts on fire regimes and key ecosystem services in Rocky Mountain forests. Forest Ecology and Management 327: 290-305.

natural fire regimes is key to evaluating forest restoration needs, since they provide a benchmark for determining the degree to which current forest conditions deviate from their "historical range of variability." Altered fire regimes, often due to long-term fire suppression, are a principle cause of elevated fire risk in many places, and re-establishing an area's natural fire regime can therefore be an important goal for forest management and restoration.

Low-intensity prescribed burn in ponderosa pine forest on the Coconino National Forest, Arizona. Photo: Brady Smith, USFS.

Beginning in the early 1900s, Federal and State agencies began carrying out aggressive fire suppression policies in efforts to reduce the loss of economically valuable timber and to protect communities. The role of fire in maintaining healthy ecosystems was not well understood at the time, and just as with the unbounded predator control taking place during that same period, these management policies produced unintended and far-reaching ecological consequences. Lack of fire in many fire-dependent forests led to the build-up of flammable materials and significantly affected forest successional patterns and processes. Changes were particularly dramatic in areas where fires historically were relatively frequent and of low intensity. Increasingly overgrown conditions and high fuel loads elevated fire risk in many areas, leaving them ripe for the ignition and spread of high intensity and severe burns.

Over the past forty years there has been an enormous amount of research on the fundamental role of fire in forest systems, resulting in an increased awareness and appreciation of fire as an essential natural process. This has been accompanied by major advances in efforts to re-establish natural fire regimes by putting fire back into these systems, as well as development of other ecological restoration techniques designed to enhance forest health and resilience.

> *Over the past forty years there* has been an enormous amount of research on the fundamental role of fire in forest systems, resulting in an increased awareness and *appreciation of fire as an essential* natural process.

Addressing the Massive Forest Restoration Backlog

The scale of forest restoration needs is enormous. The U.S. Forest Service estimates that between 65 and 82 million acres are in need of restoration just on lands within their 193 million acre national forest and grassland system.[73] There is, however, no "one size fits all" approach for forest restoration given the wide range of forest types, natural fire regimes, and current watershed conditions. Rather, restoration and hazardous fuels reduction efforts must be firmly grounded in an understanding of the dynamics of particular forest types, take into account the local and regional ecological context, as well as the needs and concerns of local communities, industry, and other stakeholders.

Collaborative forest restoration partnership in the Greater La Pine Basin, Oregon. Photo: USFS.

---

[73] U.S. Forest Service. 2015. *From Accelerating Restoration to Creating and Maintaining Resilient Landscapes and Communities Across the Nation: Update on Progress from 2012.* FS-1069. U.S. Forest Service, Department of Agriculture. https://www.fs.fed.us/sites/default/files/accelerating-restoration-update-2015-508-compliant.pdf.

Reintroducing fire into systems through the use of prescribed or controlled burns is one of the most important restoration approaches, and there is a need to dramatically expand the appropriate application of this management tool. There are places, however, where fuel loads are simply too high or conditions too dangerous for prescribed burns to be safely used. In these instances hazardous fuel loads can be reduced through a variety of mechanical thinning techniques. Ecologically appropriate thinning usually emphasizes removal of small diameter trees and dense understory vegetation. Salvage logging, which focuses on post-fire harvesting of larger-diameter standing trees, is far more controversial. Although providing economic benefit, it should only be considered on a site-specific basis since the practice can have significant ecological impacts, and its effects on reducing (or even increasing) fire risk is actively debated.[74],[75] Other forest restoration approaches include controlling invasive species, restoring streams, replacing undersized culverts, enhancing wildlife habitat, and decommissioning old forest roads.

In carrying out restoration and hazardous fuels reductions, it is also important to ensure that broader ecological and wildlife needs are taken into consideration. For example, many wildlife species depend on snags (standing dead trees) and downed woody debris for different stages in their life cycles. Some species, such as the declining black-backed woodpecker, are almost entirely dependent on post-fire snags.[76]

In an era of rapid climate change, forest restoration efforts increasingly will need to take future climatic conditions into account, rather than base management decisions on historical climatic conditions. An emphasis on climate adaptation and resilience in forest management will be especially

---

[74] Donato, D.C., J.B. Fontaine, J.L. Campbell, W.D. Robinson, J.B. Kauffman and B.E. Law. 2006. Post-wildfire logging hinders regeneration and increases fire risk. Science 311. 352.

[75] Fraver, S., T. Jain, J.B. Bradford, A.W. D'Amato, D. Kastendick, B. Palik, D. Shinneman and J. Stanovick. 2011. The efficacy of salvage logging in reducing subsequent fire severity in conifer-dominated forests of Minnesota, U.S.A. Ecological Applications. 21:1895-901.

[76] Gillis, J. 2017. Let forest fires burn? What the black-backed woodpecker knows. New York Times, August 6, 2017. https://www.nytimes.com/2017/08/06/science/let-forest-fires-burn-what-the-black-backed-woodpecker-knows.html?mcubz=0&_r=0.

important given the long life span of most tree species.[77,78] Post-fire recovery efforts, in particular, will need to carefully consider the species composition and genetic variability in plant materials used in restoration efforts in order to anticipate and prepare for changing temperature and precipitation patterns and resulting shifts in habitat suitability.

Given the enormous scale of restoration needs, it is also important to carefully target areas most in need of restoration and hazardous fuels reduction.[79] In general, fuel reduction treatments will be most appropriate close to and around wildland communities, rather than in remote backcountry and wilderness areas. Not only is the wildland-urban interface where people and property are at highest risk from wildfire, but given the preponderance of human-caused ignitions, areas of high human activity are also some of the places most likely for fires to start.[80]

In recent years there has been considerable progress in addressing the massive forest restoration need, and the Forest Service currently is treating about five million acres a year, aided by a number of promising partnership programs.[81] The Collaborative Forest Landscape Restoration Program, for example, has worked with partners to reduce wildfire risk on 1.45 million acres and improve wildlife habitat on 1.3 million acres.[82] This innovative program was created to encourage partnerships between the federal government and diverse local interests, including sawmill owners,

---

[77] Peterson, D.L., C.I. Millar, L.A. Joyce, M.J. Furniss, J.E. Halofsky, R.P. Neilson and T.L. Morelli. 2011. Responding to Climate Change in National Forests: A Guidebook for Developing Adaptation Options. General Technical Report PNW-GTR-855. U.S. Department of Agriculture, Forest Service, Pacific Northwest Research Station.

[78] Millar, C.I. and N.L. Stephenson. 2015. Temperate forest health in an era of emerging megadisturbance. Science 349: 823-826.

[79] Vaillant, N.M. and E.D. Reinhardt. 2017. An evaluation of the Forest Service hazardous fuels treatment program—are we treating enough to promote resiliency or reduce hazard? Journal of Forestry 115: 300-308.

[80] Stein, S.M., J. Menakis, M.A. Carr, S.J. Comas, S.I. Stewart, H. Cleveland, L. Bramwell and V.C. Radeloff. 2013. Wildfire, wildlands, and people: understanding and preparing for wildfire in the wildland-urban interface. General Technical Report RMRS-GTR-299, U.S. Department of Agriculture, Forest Service, Rocky Mountain Research Station.

[81] U.S. Forest Service. 2015. *From Accelerating Restoration to Creating and Maintaining Resilient Landscapes and Communities Across the Nation: Update on Progress from 2012*. FS-1069. U.S. Forest Service, Department of Agriculture. https://www.fs.fed.us/sites/default/files/accelerating-restoration-update-2015-508-compliant.pdf.

[82] Ibid.

conservationists, businesses, and sportsmen. By creating opportunities for dialogue and collaboration among groups that often have been adversaries, the program is designed to promote more science-based planning and restoration, and fewer court challenges. These creative collaborations have helped increase Forest Service timber volume by 20 percent since 2008. Other policies, such as Good Neighbor Authority and Stewardship Contracting, have also been put in place to create additional incentives for forest management and restoration, and provide significant benefits to communities and wildlife. While more work is needed, we can look to these collaborative models for ways to address the many issues facing our forests.

Restoring and enhancing the health of our forests has multiple societal benefits beyond traditional uses, such as providing timber and livestock forage. These include fueling our growing outdoor economy, providing abundant and clean water, enhancing wildlife populations, and sequestering and storing carbon. Healthy forests are essential for the outdoor recreation industry, which currently contributes $887 billion to our national economy annually, is responsible for 7.6 million direct jobs, and generates $124.5 billion in federal, state, and local tax revenue.[83] Forests are hugely important for producing the water that people depend on, and some 180 million people in over 68,000 communities rely on national forest lands to capture and filter their drinking water.[84] Forests also provide habitat for a vast array of wildlife, yet nearly one-in-five forest-dependent animal species is imperiled or vulnerable.[85] Forests are also a major factor in sequestering and storing carbon that would otherwise enter the atmosphere and contribute to climate change. Indeed, U.S. forests account for more than 90 percent of the country's carbon sink.[86]

---

[83] Outdoor Industry Association. 2017. The Outdoor Recreation Economy. https://outdoorindustry.org/wp-content/uploads/2017/04/OIA_RecEconomy_FINAL_Single.pdf.
[84] U.S. Forest Service. nd. Water facts. Accessed October 4, 2017. https://www.fs.fed.us/managing-land/national-forests-grasslands/water-facts.
[85] Heinz Center. 2008. *The State of the Nation's Ecosystems 2008*. Washington, DC: Island Press.
[86] White House Council on Environmental Quality. 2016. *United States Mid-century Strategy for Deep Decarbonization*. Washington, DC. https://unfccc.int/files/focus/long-term_strategies/application/pdf/us_mid_century_strategy.pdf.

FIXING THE FEDERAL FIRE BUDGET

Happy Dog fire, Oregon, 2017. Kari Greer, USFS.

Despite the enormous benefits that healthy forests provide, federal efforts to enhance their health and resilience are being impeded by antiquated approaches to funding the accelerating costs of fighting wildfires. Addressing the massive restoration backlog starts with fixing the broken federal fire budgeting process.

Traditionally, federal budgets for fighting wildfire have been based on rolling ten year historical averages. While this budgeting approach may have worked in prior eras, given the dramatic increase in the number and size of wildfires, and the rapidly escalating costs of fighting these fires, this retrospective approach is clearly inadequate for today's needs. Today, fighting wildfire consumes more than 50 percent of the Forest Service's budget, and this number could grow to 67 percent over the next decade, in large part due to a changing climate and ensuing drought conditions.[87] Annually, the United States spends more than $1 billion to fight wildfires,[88] and in 2015, one of the worst years for wildfires on record, the U. S. government spent more than $2.6 billion.[89] Large catastrophic megafires eat up a disproportionate amount of federal resources, with just one to two percent of fires consuming 30 percent or more of agency firefighting budgets each year.[90] These costs are only expected to rise as the climate continues to change, and as more homes are built in fire-prone wildland areas.

Unfortunately, the current fire funding system is not only inadequate for fighting fires, it is also compromising the ability of agencies to proactively reduce fire risks. Because of the retrospective budgeting process for fire control and suppression funds, during active fire years (which are now the norm) as fire season progresses and available wildfire funds are exhausted, money is shifted (euphemistically termed "borrowed") from accounts funding other important agency activities. This fire borrowing severely affects the government's ability to carry out the programs for which those funds were originally intended, including recreation, wildlife management, and forest restoration. A permanent fix is needed to address this growing problem.

Part of that fix must include a recognition that the number, size, and cost of wildfires on federal lands is increasing and should be budgeted for appropriately. Additionally, instead of expecting agencies to cover the exceptional costs of fighting truly catastrophic fires from their limited annual appropriations, these expenses should be eligible to be covered by specially appropriated disaster funds, similar to how hurricanes, tornadoes, and major flood events have long been treated.

---

[87] U.S. Forest Service. 2015. *The Rising Cost of Wildfire Operations: Effects on the Forest Service's Non-Fire work.* U.S. Forest Service, Department of Agriculture. https://www.fs.fed.us/sites/default/files/2015-Fire-Budget-Report.pdf.

[88] National Interagency Fire Center. 2017. Federal Firefighting Costs. Accessed October 4, 2017. https://www.nifc.gov/fireInfo/fireInfo_documents/SuppCosts.pdf.

[89] Streater, S. 2016. Policymakers must prepare for larger blazes – report. Greenwire, January 18, 2016. https://www.eenews.net/greenwire/stories/1060030747.

[90] U.S. Forest Service. 2015. *The Rising Cost of Wildfire Operations: Effects on the Forest Service's Non-Fire Work.* U.S. Department of Agriculture. https://www.fs.fed.us/sites/default/files/2015-Fire-Budget-Report.pdf.

Recommendations

The U.S. must address the growing threat of megafires through a comprehensive wildfire funding fix, through dramatically scaling up the pace of forest restoration, through incorporating climate adaptation practices into forest restoration and management, and through achieving significant reductions in climate-altering carbon pollution. By restoring and better managing U.S. forests, it is possible to reduce fire risks to communities, increase populations of cherished wildlife species, and protect our climate by enhancing the carbon sequestration and storage potential of our forests. These steps will help ensure that America's forests will be sustainable and resilient in the face of a rapidly changing and uncertain future, and will be capable of continuing to provide important economic, ecological, and societal benefits.

Ensure Adequate and Dependable Wildfire Funding
- Provide sufficient funding for federal agencies to respond to wildfires, recognizing the growing average annual cost of firefighting;
- Allow the Forest Service to access a disaster funding account for catastrophic and extraordinarily costly fires;

Prescribed burn, western Oregon. Photo: BLM.

- End the transfer of funds from conservation, forest management, and other non-fire programs to cover growing fire suppression costs; and
- Increase federal funding to implement proactive restoration and fire risk reduction, such as prescribed burns (when appropriate), on public and private forested lands.

Accelerate Restoration Projects and Improve Forest Management
- Prioritize restoration projects that achieve outcomes related to improved forest resilience, increased wildlife populations, and watershed health;
- Carry out significantly more prescribed and managed burns in fire-adapted forests;
- Improve environmental review processes for large-scale landscapes and for project-level sites in the short-term that help achieve and facilitate forest restoration and healthy forest management; and
- Expand on successful nationwide policies such as Good Neighbor Authority and Stewardship Contracting, which provide models for how stakeholders can work collaboratively towards mutually beneficial forest management goals and bring additional non-federal resources to restoration projects.

Prepare for Changing Climatic Conditions and for More Frequent and Severe Wildfires
- Integrate climate adaptation principles into forest management and restoration efforts to ensure they are designed for future, rather than past, climatic conditions and promote sustainability and resilience of forest resources;
- Encourage wildland communities to incentivize new housing in areas of lower fire risk, promote or mandate the use fire-resistant building materials, and adopt other fire-wise approaches for reducing wildfire risk; and

- Prioritize hazardous fuel reduction investments in the wildland-urban interface where they will have the greatest effect on reducing the impacts and costs of wildfires.

Reduce Climate-Altering Carbon Pollution
- Manage forests on federal lands in ways that promote continued capture and storage of carbon, and foster financial incentives and markets to encourage carbon sequestration on private forest lands;
- Implement common sense safeguards—like greenhouse gas limits for power plants, vehicles, and oil and gas facilities—that are needed to protect public health and wildlife from climate impacts;

Post-fire wildflower display. Photo: Damian Gadal.

- Enact measures at state and local levels of government to curb carbon pollution and expand use of clean, renewable energy; and
- Ensure that federal agencies have sufficient resources to pursue important climate change research and monitoring, and to spur the development and adoption of clean energy technologies.

***

Greg Walden,
Oregon Chairman

Frank Pallone, Jr.,
New Jersey Ranking Member

One Hundred fifteenth Congress
Congress of the United States
House of Representatives
Committee on Energy and Commerce

October 9, 2018

The Honorable Herman E. Baertschiger, Jr.
Senator
Oregon State Senate
900 Court Street, N.E.
Salem, OR 97301

Dear Senator Baertschiger:

Thank you for appearing before the Subcommittee on Environment on September 13, 2018, to testify at the hearing entitled "Air Quality Impacts of Wildfires: Mitigation and Management Strategies."

Pursuant to the Rules of the Committee on Energy and Commerce, the hearing record remains open for ten business days to permit Members to submit additional questions for the record, which are attached. To facilitate the printing of the hearing record, please respond to these questions with a transmittal letter by the close of business on Tuesday, October 23, 2018. Your responses should be mailed to Kelly Collins, Legislative Clerk, Committee on Energy and Commerce, 2125 Rayburn House Office Building, Washington, DC 20515 and e-mailed in Word format to kelly.collins@mail.house.gov.

Thank you again for your time and effort preparing and delivering testimony before the Subcommittee.

Sincerely, John Shimkus
Chairman
Subcommittee on Environment

cc: The Honorable Paul Tonko, Ranking Member, Subcommittee on Environment
Attachment

Attachment-Additional Questions for the Record

The Honorable John Shimkus

1) What is necessary to increase the pace and scale of prescribed burning and other active forest management activities? More specifically what needs to happen at the Federal level vs State and local levels?
2) Can you provide your perspective on whether more coordination among federal and state authorities is needed to make a meaningful difference in reducing the risks of catastrophic wildfires?

3) Should air quality considerations play a greater role in informing decisions related to wildfire suppression and forestry management planning, and if so, how so?

RESPONSES BY OREGON STATE SENATOR HERMAN E. BAERTSCHIGER, JR.

To: The Honorable John Shimkus

1. What is necessary to increase the pace and scale of prescribed burning and other active forest management activities? More specifically what needs to happen at the Federal level vs State and local levels?

To increase the pace and scale of active forest management activities, uninterrupted investment by the Congress is necessary. Continuity and consistency with long-term support is crucial to the success of any effort related to active forest management. The solution to Western forest management is very long-term. It should be at least a 100 year plan to make an impact. Long term goals are difficult for most people to conceive, but is the reality in forest management. Forest management is dynamic {an ever changing environment} where one must adjust strategies, goals and tactics to meet the ever changing events. A consistent set of principles and directives that are adequately funded for the very long-term is needed to have successful active forest management.

Coordination between Federal, State and local agencies must include common and coordinated objectives and goals. This coordination is also required to increase the pace and scale of active forest management. Most states have a fire suppression model for state protected lands. The Federal government for at least the last 20 years has operated under a fire management model. This Suppression versus Management model is unsustainable. Managing fire during peak fire season is unacceptable because of the adverse effects to the health and welfare of the people. To have successful fire and forest management there should be coordinated strategies and goals, between the Federal and State governments, especially during peak fire season. Their fire policies should be parallel and efforts should be consistent relating to same goals and outcomes by both State and Federal agencies.

2. Can you provide your perspective on whether more coordination among federal and state authorities needed to make meaningful difference in reducing the risks of catastrophic wildfires?

Managing fire during peak fire season is unacceptable because of the adverse effects to the health and welfare of the people. To have successful fire and forest management there should be coordinated strategies and goals, between the Federal and State governments, especially during peak fire season.

From a prevention and awareness perspective, identifying defensible space around communities and creating adequate escape routes that the public can be made aware of is a risk management tool that can be used to minimize the effects of catastrophic wildfires in rural communities. Fuel reduction around communities in fire prone areas can be employed to create and improve defensible space. Identification and enhanced public knowledge of escape routes in fire prone communities is an awareness tool. Much in the way that tsunami prone areas have signs and placards giving direction and creating awareness, a similar approach in fire prone areas could be used.

3. Should air quality considerations play a greater role in informing decisions related to wildfire suppression and forestry management planning, and if so, how so?

Air Quality considerations must play a greater role in decisions related to wildfire suppression and forestry management planning. Attempting to manage fire rather than suppress it during peak fire season is unacceptable because of air quality considerations. Managing fire during peak fire season is unacceptable because of the adverse effects to the health and welfare of the people. Forestry management planning, including harvest, thinning and controlled burning, can be managed correctly outside of peak fire season while still preserving air quality.

Harvest and thinning do not create significant air quality impacts. Controlled burning can be managed outside of peak fire season and the impacts of smoke can be minimized. Air Quality impacts during some of the most recent fire seasons has often exceeded hazardous levels. Therefore, a full suppression policy should be followed during peak fire season and an aggressive control burning policy for the off season to minimize the impacts of smoke, and the health risks it presents to the public.

***

Greg Walden,
Oregon Chairman

Frank Pallone, Jr.,
New Jersey Ranking Member

One Hundred fifteenth Congress
Congress of the United States
House of Representatives
Committee on Energy and Commerce

October 9, 2018

Ms. Mary Anderson
Mobile and Area Source Program Manager
Air Quality Division
Idaho Department of Environmental Quality
1410 North Hilton, Boise, ID 83706
Dear Ms. Anderson:

Thank you for appearing before the Subcommittee on Environment on September 13, 2018, to testify at the hearing entitled "Air Quality Impacts of Wildfires: Mitigation and Management Strategies."

Pursuant to the Rules of the Committee on Energy and Commerce, the hearing record remains open for ten business days to permit Members to submit additional questions for the record, which are attached. To facilitate the printing of the hearing record, please respond to these questions with a transmittal letter by the close of business on Tuesday, October 23, 2018. Your responses should be mailed to Kelly Collins, Legislative Clerk, Committee on Energy and Commerce, 2125 Rayburn House Office Building, Washington, DC 20515 and e-mailed in Word format to kelly.collins(a),mail.house.gov.

Thank you again for your time and effort preparing and delivering testimony before the Subcommittee.

Sincerely, John Shimkus
Chairman
Subcommittee on Environment

cc: The Honorable Paul Tonko, Ranking Member, Subcommittee on Environment
Attachment

ATTACHMENT-ADDITIONAL QUESTIONS FOR THE RECORD

The Honorable John Shimkus

1) What is necessary to increase the pace and scale of prescribed burning and other active forest management activities? More specifically what needs to happen at the Federal level vs State and local levels?

2) Can you provide your perspective on whether more coordination among federal and state authorities is needed to make a meaningful difference in reducing the risks of catastrophic wildfires?
3) Should air quality considerations play a greater role in informing decisions related to wildfire suppression and forestry management planning, and if so, how so?

The Honorable Richard Hudson

1) In North Carolina we recently saw one of the worst wildfires our state has seen, claiming over 55,000 acres. Not only do these blazes destroy our homes and lands, but they also impact our health. \What type of risk communication strategy should states like mine who normally don't experience major wildfires put out to info on the public of the risks associated with wildfires?
2) From a health perspective, what are the impacts of wildfires? More specifically, how does smoke impact sensitive populations such as the elderly, children, or individuals who suffer from respiratory challenges?
    a. How do you work with public health officials to alert communities about the health impacts of wildfire smoke? What steps can individuals and communities take to minimize the health impacts of wildfire smoke?

***

State of Idaho
Department of Environmental Quality

October 23, 2018
Ms. Kelly Collins
Legislative Clerk
Committee on Energy and Commerce

2125 Rayburn House Office Building
Washington, DC 20515

RE: Responses to Questions for the Record on "Air Quality Impacts of Wildfires: Mitigation and Management Strategies."

Dear Ms. Collins:

Thank you for the opportunity to testify at the hearing entitled "Air Quality Impacts of Wildfires: Mitigation and Management Strategies" on Thursday, September 13, 2018. Per your request, please find attached my responses to your additional questions for the record.

If you would like further clarification regarding any of the answers provided, please don't hesitate to contact me.

Sincerely,
Mary Anderson
Mobile and Area Source Program Manager
Attachment (1)

Responses to Questions for the Record for Mary Anderson, Idaho Department of Environmental Quality, Hearing: Air Quality Impacts of Wildfires: Mitigation and Management Strategies

The Honorable John Shimkus

1. What is necessary to increase the pace and scale of prescribed burning and other active forest management activities? More specifically what needs to happen at the Federal level vs State and local levels?

Answer: From an air quality regulatory agency perspective, more collaboration and coordination is needed before more prescribed burning can be achieved. State and federal land managers need to coordinate closely with Idaho DEQ and other private burners. In order for the prescribed burning program to be successful, there also needs to be a collaboration and communication with the public. This coordination and collaboration needs to happen long before the prescribed burn is scheduled to occur. The following steps need to be accomplished prior to increasing the use of prescribed burning:

- Establish airshed groups – These groups will be based on defined geographic area and include all burners as well as other stakeholders such as the public. These groups will help prioritize burns and educate the public on the need for prescribed burning.
- Develop a comprehensive smoke management plan – Because many types of burning occur in Idaho, a plan that addresses all open burning is needed to ensure public health is protected.
- Coordination across state lines – Similar to the MT/ID Airshed group, burners and air quality agencies need to coordinate burn decisions across state lines.
- Increase in staffing – As I stated in my testimony, the current staffing level can barely keep up with the current level of burning. Additional staffing would be needed to ensure the public health is protected while maximizing the opportunities to burn.
- Evaluate all possible solutions – Prescribed burning cannot be seen as the only solution to the wildfire problem, other forest management techniques to remove wildland fuels must become part of the solution.
- Use smoke management principles – It is important to remember that reasonable and effective smoke management principles and decisions must be used when conducting prescribed burning to truly lessen smoke impacts and not simply move smoke from one time of year to another. The addition of more smoke into some of Idaho's airsheds through increasing the use of prescribed fire during October

through December will put some communities in jeopardy of exceeding the national ambient air quality standards for PM2.5 as well as adversely impacting the public's health. Data can be flagged as "exceptional," thereby excluding it from attainment demonstrations, but only if adequate smoke management principles are adopted and applied.
- Reducing emissions – Incentivizing the use of alternatives to open burning of wildland fuels as well as the replacement of old inefficient woodstoves are both needed concurrently to reduce the amount of emissions competing for the airshed.

2. Can you provide your perspective on whether more coordination among federal and state authorities is needed to make a meaningful difference in reducing the risks of catastrophic wildfires?

Answer: From my perspective there is always room for more coordination. Managing lands involves many agencies at both the state and federal level. The decisions made by these agencies also impact the public that live near these lands. With multiple agencies and stakeholders, collaboration is also needed to truly make a difference. More coordination will only be effective if all participants value each other's missions and mandates and are committed to finding common goals and balanced solutions.

3. Should air quality considerations play a greater role in informing decisions related to wildfire suppression and forestry management planning, and if so, how so?

Answer: Yes, air quality should play a greater role in informing decisions related to wildfire suppression and forestry management planning. One of the guiding principles of the 1995 Federal Fire Policy (reaffirmed in 2001 and 2009) is that "Fire management plans and activities incorporate public health and environmental quality considerations." When responding to wildfire, air quality should be one of the risk factors evaluated when determining the appropriate response, whether it is direct attack, conducting a back burn, or establishing point protection. When air quality deteriorates to the unhealthy, very unhealthy, and hazardous conditions, air quality should become a more important component of the decision making process. The same focus of wildfire suppression to commit resources for the protection of structures should be afforded the protection to public health whenever possible.

The Honorable Richard Hudson

1. In North Carolina we recently saw one of the worst wildfires our state has seen, claiming over 55,000 acres. Not only do these blazes destroy our homes and lands, but they also impact our health. What type of risk communication strategy should states like mine who normally don't experience major wildfires put out to inform the public of the risks associated with wildfires?

Answer: The key to responding to wildfire impacts is preparing ahead. Identifying key stakeholders and developing outreach material on the fly during the event is not effective. I recommend working during the winter months to develop a communication strategy with the stakeholders in your state. Idaho relies heavily on the Wildfire Smoke A Guide for Public Health Officials to make recommendations to the public (htttps://www3.epa.gov/airnow/wildfire_may2016.pdf). We also work closely with our neighboring states when developing smoke forecasts and communicating health recommendations for the public. To improve coordination and collaboration during wildfire response, DEQ and several partners developed a wildfire smoke response protocol to help mitigate impacts on public health by guiding air quality information distribution and health related messaging and outreach responses through the appropriate agency. Idaho highly recommends developing a communication strategy with your state's health partners.

2. From a health perspective, what are the impacts of wildfires? More specifically, how does smoke impact sensitive populations such as the elderly, children, or individuals who suffer from respiratory challenges?

Answer: When it comes to smoke impacts, the sensitive populations include children (especially children age 7 and younger), pregnant women, older adults (over 65 years of age), individuals with pre-existing lung and cardiovascular conditions, and smokers. Symptoms from smoke inhalation can include shortness of breath, chest pain or tightness, headaches, coughing, irritated sinuses, stinging eyes, sore throat, and fatigue.

   a. How do you work with public health officials to alert communities about the health impacts of wildfire smoke? What steps can individuals and communities take to minimize the health impacts of wildfire smoke?

Answer: Idaho DEQ coordinates closely with the Idaho Department of Health and Welfare (IDHW) as well as the local Health Districts during a wildfire smoke event. In collaboration with many stakeholders, we developed a Wildfire Smoke Response Protocol that identifies key stakeholders and contact information, agency expertise, triggers for agency actions, and recommended actions. DEQ and Tribal Air quality agencies provide current and forecasted air quality while the IDHW and Health Districts communicate health related information and recommendations to the public. The US Forest Service is also part of this collaboration and provides insight into current wildfire behavior to the group. We follow the information presented in the Wildfire Smoke A Guide for Public Health Officials to make recommendations to the public (https://www3.epa.gov/airnow/wildfire_may2016.pdf). Individuals and communities can take the following steps to minimize health impacts:

- Stay indoors
  - Keep windows closed and run a filtered air conditioner with the fresh air intake closed
  - Use room high-efficiency particulate air (HEPA) cleaners that DO NOT produce excess ozone.
  - Create a clean room at home
- Reduce activity
- Stay well-hydrated by drinking plenty of water
- Follow your doctor's advice about medication and your respiratory management plan if you have asthma or another lung disease. This is best done prior to wildfire smoke impacting the community. Plan ahead.
- Switch to eyeglasses if you wear contacts
- Do not add to indoor pollution – avoid frying or broiling when cooking, do not vacuum or smoke
- Do not add to outdoor pollution – do not burn wood, limit using gas lawnmowers and driving

***

Greg Walden,
Oregon Chairman

Frank Pallone, Jr.,
New Jersey Ranking Member

One Hundred fifteenth Congress
Congress of the United States
House of Representatives
Committee on Energy and Commerce

October 9, 2018
Ms. Sonya Germann
State Forester

Forestry Division
Montana Department of Natural Resources and Conservation
2705 Spurgin Road
Missoula, MT 59804

Dear Ms. Germann:

Thank you for appearing before the Subcommittee on Environment on September 13, 2018, to testify at the hearing entitled "Air Quality Impacts of Wildfires: Mitigation and Management Strategies."

Pursuant to the Rules of the Committee on Energy and Commerce, the hearing record remains open for ten business days to permit Members to submit additional questions for the record, which are attached. To facilitate the printing of the hearing record, please respond to these questions with a transmittal letter by the close of business on Tuesday, October 23, 2018. Your responses should be mailed to Kelly Collins, Legislative Clerk, Committee on Energy and Commerce, 2125 Rayburn House Office Building, Washington, DC 20515 and e-mailed in Word format to kelly.collins@mail.house.gov.

Thank you again for your time and effort preparing and delivering testimony before the Subcommittee.

Sincerely, John Shimkus
Chairman
Subcommittee on Environment

cc: The Honorable Paul Tonko, Ranking Member, Subcommittee on Environment

Attachment

ATTACHMENT-ADDITIONAL QUESTIONS FOR THE RECORD

The Honorable John Shimkus

1) What is necessary to increase the pace and scale of prescribed burning and other active forest management activities? More specifically what needs to happen at the Federal level vs State and local levels?
2) Can you provide your perspective on whether more coordination among federal and state authorities is needed to make a meaningful difference in reducing the risks of catastrophic wildfires?
3) Should air quality considerations play a greater role in informing decisions related to wildfire suppression and forestry management planning, and if so, how so?

October 23, 2018

RESPONSES TO QUESTIONS FOR THE RECORD FOR SONYA GERMANN, MONTANA STATE FORESTER HEARING: AIR QUALITY IMPACTS OF WILDFIRES: MITIGATION AND MANAGEMENT STRATEGIES (SEPT 13, 2018)

Questions from the Honorable John Shimkus

1. What is necessary to increase the pace and scale of prescribed burning and other active forest management activities? More specifically what needs to happen at the Federal level vs State and local levels?

The USDA Forest Service State & Private Forestry programs support essential investments in the health and management of our nation's forests. Any increase in federal appropriations to those programs will be critical in helping reduce hazardous fuel loads on private lands in and around the Wildland Urban Interface, reducing the risk of wildfire for communities. Additionally, the Landscape Scale Restoration (LSR) program is a key priority for State Foresters, allowing land managers to address national priorities identified in state Forest Action Plans. Codifying this program and funding it appropriately would allow state forestry agencies to tackle the nation's most pressing forest priorities in the most cost-effective, collaborative, and coordinated ways.

Increasing the implementation of authorities granted by Congress, such as the Good Neighbor Authority and categorical exclusions for wildfire resilience projects, will ensure that high priority areas are actively managed. The recent passage of the wildfire funding solution that will end the detrimental practice of fire borrowing will assist the Forest Service in utilizing these available tools. Montana, like other states, appreciates Congress' support for increasing the pace and scale of forest restoration and for the authorities that facilitate strong cooperation between the state and federal agencies. We would urge Congress to consider additional actions, such as the recommendations cited in the Western Governors' Association National Forest and Rangeland Management Initiative to carryout hazardous fuels reduction and support advancements in the use of prescribed fire. This initiative represents a multi-state, bipartisan collaborative perspective on promoting health and resilience of forests and rangelands in the West and highlights mechanisms to bring states, federal land managers, private landowners, and stakeholders together to discuss issues and opportunities in forest and rangeland restoration and management emphasizing investments in all lands/cross-boundary management opportunities.

Additionally, we would urge Congress to continue increasing its support of collaborative efforts on federal forest land management. In Montana, these groups have done the hard work of reaching agreement on intractable land management issues, and their continued engagement is critical to the success of increasing the pace and scale of forest management within the state.

2. Can you provide your perspective on whether more coordination among federal and state authorities is needed to make a meaningful difference in reducing the risks of catastrophic wildfires?

Increased coordination is critical to successful forest management across shared boundaries. Thankfully this concept is widely recognized by federal and state land management agencies as a method to success in making meaningful differences across the landscape. Right now there is a unique opportunity for an all-hands, all-lands approach where federal, state, and local governments as well as collaborative groups are all involved in the planning process as well as being on the front lines. Additionally, authorities such as the Good Neighbor Authority support opportunities to increase cross-boundary coordination between states and federal partners in order to accomplish more restoration work.

The recent initiative from the USDA Forest Service, *"Toward Shared Stewardship Across Landscapes: An Outcome Based Investment Strategy,"* also calls for increasing coordination. This strategy highlights the Forest Service's vision of bringing the States together with the Agency and other partners to identify priority areas for increasing active forest management. Since the establishment of the Wildland Fire Leadership Council in 2002, there has been an ongoing effort by federal, state, local governments, and collaborative groups to increase coordination around the efforts to reduce the risk of catastrophic fire. *The National Cohesive Strategy* is an example of this improved coordination that continues today.

3. Should air quality considerations play a greater role in informing decisions related to wildfire suppression and forestry management planning, and if so, how so?

Air Quality considerations could serve a greater role in informing decisions related to wildfire suppression and forestry management planning.

We cannot prevent wildfire, but we can influence the way that wildfire burns. We can also work to lessen the hazardous fuel loads on the ground through mechanical thinning and prescribed fire, which will mitigate the amount of smoke communities experience as a result of wildfire. The data shows that in Montana, over an 11-year period, air quality standards were surpassed 579 times due to wildfire, while air quality standards were surpassed only 4 times due to prescribed fire. In both the updating of the National Ambient Air Quality Standard (NAAQS) for PM 2.5 (81 CFR 164, pg. 58010) and the updating of the Exceptional Events Rule (81 CFR 191, pg. 68216), the Environmental Protection Agency (EPA) clearly documents the role of wildfire as an emissions source and the relevance of prescribed fire use and fuels management to reduce the risk of catastrophic wildfire. It is becoming increasingly evident through both research and experience that without prescribed fire and the relatively small amount of managed smoke that comes with it, we are perpetuating the conditions that generate catastrophic fires and resulting air quality issues, while simultaneously putting people and their communities at risk. NASF, my fellow State Foresters and I, are ready to work with members of the House Energy and Commerce Committee and staff, to examine possible legislative solutions to allow for more implementation of prescribed burns and mechanical thinning.

***

Greg Walden,
Oregon Chairman

Frank Pallone, Jr.,
New Jersey Ranking Member

One Hundred fifteenth Congress
Congress of the United States
House of Representatives
Committee on Energy and Commerce
October 9, 2018

Mr. Collin O'Mara
President and CEO
National Wildlife Federation
1200 G Street; Suite 900
Washington, DC 20005

Dear Mr. O'Mara:

Thank you for appearing before the Subcommittee on Environment on September 13, 2018, to testify at the hearing entitled "Air Quality Impacts of Wildfires: Mitigation and Management Strategies."

Pursuant to the Rules of the Committee on Energy and Commerce, the hearing record remains open for ten business days to permit Members to submit additional questions for the record, which are attached. To facilitate the printing of the hearing record, please respond to these questions with a transmittal letter by the close of business on Tuesday, October 23, 2018. Your responses should be mailed to Kelly Collins, Legislative Clerk, Committee on Energy and Commerce, 2125 Rayburn House Office Building, Washington, DC 20515 and e-mailed in Word format to kelly.collins@mail.house.gov.

Thank you again for your time and effort preparing and delivering testimony before the Subcommittee.

Sincerely, John Shimkus
Chairman
Subcommittee on Environment

cc: The Honorable Paul Tonko, Ranking Member, Subcommittee on Environment

Attachment

ATTACHMENT-ADDITIONAL QUESTIONS FOR THE RECORD

The Honorable John Shimkus

1) What is necessary to increase the pace and scale of prescribed burning and other active forest management activities? More specifically what needs to happen at the Federal level vs State and local levels?
2) Can you provide your perspective on whether more coordination among federal and state authorities is needed to make a meaningful difference in reducing the risks of catastrophic wildfires?
3) Should air quality considerations play a greater role in informing decisions related to wildfire suppression and forestry management planning, and if so, how so?

RESPONSE FROM COLLIN O'MARA TO THE HOUSE COMMITTEE ON ENERGY AND COMMERCE SUBCOMMITTEE ON ENVIRONMENT, HEARING ENTITLED "AIR QUALITY IMPACTS OF WILDFIRES: MITIGATION AND MANAGEMENT STRATEGIES," SEPTEMBER 13, 2018

Thank you again for the opportunity to testify, and for the additional questions you have posed. My answers are contained below.

1. What is necessary to increase the pace and scale of prescribed burning and other active forest management activities? More specifically what needs to happen at the Federal level vs State and local levels?

Funding, collaboration, and focus. Earlier this year Congress passed the Fire Fix that will ensure that, starting in fiscal year 2020, the Forest Service will no longer have to extinguish all resources from its forest management accounts to cover the increasing costs of fighting the larger, more numerous, and more intense forest fires being fueled by climate change. This was an extremely important step that Congress took and we thank all of the Members of this Committee who supported it and worked to enact it.

That said, we are extremely frustrated that the Fire Fix will not take effect until FY20, when it should have begun in the second half of FY18. Given the severity of fires we experienced in FY18 and what we're expecting for FY19, we urge this Committee to work with other committees of jurisdiction to begin the fire funding fix this year (FY19), rather than waiting for yet another year to begin restoration at-scale, while local communities suffer even more devastating health and economic impacts.

Further, the fire funding fix in and of itself is not enough. After years of starving the Forest Service budget combined with its own restoration budget cannibalization to fight fires, we urge Congress to follow through by making sure the Forest Service has sufficient funding on an annual basis to restore our National Forests to health and resilience, through forwardthinking restoration projects, prescribed burns, and other efforts. Such funding should at least be commensurate with the new wildfire disaster funds to be allocated for fighting wildfires, so we are reducing the long-term restoration deficit. Further, in out years, Congress should ensure that the Forest Service is spending the resources freed up by the fire funding fix on the restoration, reforestation, and proactive management.

More collaboration by the Forest Service with more partners will also increase the pace and scale of forest restoration projects. The more buy-in and local, regional, and national support the Forest Service has for its work, the more likely those projects are to move forward without controversy or pushback. The Forest Service's Shared Stewardship strategy released earlier this year sets it on the right course toward increased coordination with the full range of national forest stakeholders and the public. This strategy seeks to address fire at scale, and emphasizes and prioritized greater coordination with states in particular, and with other stakeholders. Congress should support the agency in implementing the shared strategy and provide funding as well as oversight to make sure the Forest Service stays on track and delivers forest restoration results.

A focus on restoration results will also be needed to increase the number of prescribed burns and other forest restoration projects. The Forest Service has plenty of policy and programmatic tools to restore the national forests, and now it has more funding available for this purpose thanks to Congressional action earlier this year. Some additional policy tools or flexibility from Congress might help around the edges but are not a priority right now, when the Forest Service has barely begun to implement the tools Congress recently provided this spring as part of the fire funding fix. Instead, the agency must make sure all of its forest restoration tools and authorities, and the restoration projects it is implementing on-the-ground through them, are actually focused on improving the resilience of forest ecosystems and on improving wildlife habitat and delivering results. The restoration work these authorities were set up to achieve is far too important, and the scale of restoration needed far too great, to allow any projects under these programs to focus on anything but restoring the health and resilience of our national forests. Yet there is a tendency for the Forest Service to prioritize generating receipts from the sale of commercially viable timber to cover costs over delivering restoration results. A commercial timber program is a legitimate and important use of the national forests, but seeking receipts through forest restoration programs and projects serves as a distracting and corrupting influence on those programs.

Receipts from the sale of timber must be a byproduct of restoration projects, not an objective. The Forest Service therefore needs to measure the success of its forest restoration program in terms of community fire risk reduction, forest resilience, wildlife habitat, carbon storage, water quality, and other measurable results on the ground. Congress can help by providing oversight to make sure objective forest restoration results are being delivered irrespective of receipts generated. For example, Congress should make sure the restoration authorities it has provided the agency such as the Collaborative Forest Landscape Restoration Program, Stewardship Contracting, and Good Neighbor Authority are fulfilling their restoration purposes and not authorizing commercial timber sales except as necessary to achieve a specific restoration result.

2. Can you provide your perspective on whether more coordination among federal and state authorities is needed to make a meaningful difference in reducing the risks of catastrophic wildfires?

More—earlier and more strategic—coordination among federal and state authorities will lead to a meaningful difference in reducing the risk of catastrophic wildfires. The key will be to make sure the coordination is centered on the restoration of forest resilience, for example through prescribed burns, the removal of flammable understory, or the retention of mature fire resistant trees. Some steps the Forest Service could take to improve coordination with states include providing more oversight of the Good Neighbor Authority program to make sure projects under this authority are prioritizing restoration over receipts, and adopting policies or regulations making sure Stewardship Contracts are only signed for legitimate forest restoration projects.

3. Should air quality considerations play a greater role in informing decisions related to wildfire suppression and forestry management planning, and if so, how so?

Air quality considerations need to play a greater role in informing forest restoration projects. As I mentioned in my testimony, prescribed burns emit 10 times to 100 times less particulate matter than typical wildfires—and prescribed burns are one of the most important forest restoration and fire risk reduction strategies. The USFS and States already take precautions to reduce the health impacts from prescribed burns, such as establishing hourly and daily PM 2.5 limits, as well as specific plans to support at-risk populations. These plans should continue to be improved to minimize further any potential health impacts from prescribed burns, but potential impacts should always be compared to the health consequences of inaction, which are often orders of magnitude greater as residents across the Northwest have experienced.

There are administrative actions that EPA can take to remove some of the disincentives for prescribed burns through both changes to implementation guidance and policy. The most important change is to eliminate the perverse incentive whereby emissions from prescribed burns ("anthropogenic ignition") are included in the calculations to determine whether a state is in attainment of the National Ambient Air Quality Standards, but wildfires ("natural ignition") are regularly excluded, despite typically emitting 90-99% more pollution. In other words, states that proactively utilize prescribed burns to reduce risks of megafires often need to find further reductions elsewhere in their state economy to offset the emissions from the prescribed burns; whereas a state that does not take such preventative action is not held accountable for the emissions from a megafire that could have been mitigated. Instead, we believe that EPA should account for all emissions and prioritize the granting of wildfire accounting exceptions to those states and communities who have ecologically-sound and landscape-scale fire programs and in doing so encouraging states to prioritize forest restoration, including prescribed burns, as a means of reducing overall emissions and adverse impacts to public health. We would glad to work with the Committee on this effort.

***

Greg Walden,
Oregon Chairman

Frank Pallone, Jr.,
New Jersey Ranking Member

One Hundred fifteenth Congress
Congress of the United States
House of Representatives
Committee on Energy and Commerce

October 9, 2018

Mr. Tom Boggus
State Forester and Director
Texas A&M Forest Service
200 Technology Way; Suite 1281
College Station, TX 77845

Dear Mr. Boggus:

Thank you for appearing before the Subcommittee on Environment on September 13, 2018, to testify at the hearing entitled "Air Quality Impacts of Wildfires: Mitigation and Management Strategies."

Pursuant to the Rules of the Committee on Energy and Commerce, the hearing record remains open for ten business days to permit Members to submit additional questions for the record, which are attached. To facilitate the printing of the hearing record, please respond to these questions with a transmittal letter by the close of business on Tuesday, October 23, 2018. Your responses should be mailed to Kelly Collins, Legislative Clerk, Committee on Energy and Commerce, 2125 Rayburn House Office Building, Washington, DC 20515 and e-mailed in Word format to kelly.collins@mail.house.gov.

Thank you again for your time and effort preparing and delivering testimony before the Subcommittee.

Sincerely, John Shimkus
Chairman
Subcommittee on Environment

cc: The Honorable Paul Tonko, Ranking Member, Subcommittee on Environment

Attachment

ATTACHMENT ADDITIONAL QUESTIONS FOR THE RECORD

The Honorable John Shimkus

1) What is necessary to increase the pace and scale of prescribed burning and other active forest management activities? More specifically what needs to happen at the Federal level vs State and local levels?
2) Can you provide your perspective on whether more coordination among federal and state authorities is needed to make a meaningful difference in reducing the risks of catastrophic wildfires?
3) Should air quality considerations play a greater role in informing decisions related to wildfire suppression and forestry management planning, and if so, how so?

The Honorable Bill Flores

1) Your testimony mentions both prescribed burning and tree thinning as important tools available to make our forests healthier and less fire prone.
   a) As a forester, could you explain how cutting trees makes the remaining forests healthier?
   b) How successful has tree thinning been in Texas and the rest of the country? Should we be doing more?

The Honorable Richard Hudson

1) In your testimony you say that "our forests are currently more fire-prone than ever." Why is this the case? How did we get to the point

where our forests are at such high risk of fire? Are there any regulations inhibiting the ability to lower our forests risks of fire?

October 23, 2018

Responses to Questions for the Record for Tom Boggus, Texas State Forester Hearing: Air Quality Impacts of Wildfires: Mitigation and Management Strategies (Sept 13, 2018)

Questions from the Honorable John Shimkus

1. What is necessary to increase the pace and scale of prescribed burning and other active forest management activities? More specifically what needs to happen at the Federal level vs State and local levels?

Increasing both prescribed burning and active forest management are critical to improving forest health and supporting economic prosperity and safety in rural communities across the country. At the state and local levels, it is imperative that we help communities and citizens understand the dire risks from catastrophic wildfire, and the tie between prescribed burning and active management to reducing that risk. Maintaining the social license to burn and harvest has always been a key to our successful forest management in Texas and many other parts of the South, but it is becoming increasing challenging as our country is urbanizing. Increasing awareness within the regulatory community is also important, helping state environmental authorities understand the benefit to allowing some prescribed fire smoke in lieu of an egregious amount of wildfire smoke and potentially catastrophic tree and property losses sometime down the road.

At the federal level, public land management agencies need more tools and resources to facilitate and streamline active forest management. Our federal forests are in dire need of restoration and fire risk reduction, and it is very much in the public interest to find ways to support getting that work done. The Good Neighbor Authority is particularly helpful, as it allows state agencies to carry out work on federal lands. Texas was an early adopter of this authority and continues to find ways to help our federal partners restore the health of our federal forests. Federal programs that support technical assistance to private landowners through the state forestry agencies are also critical to keeping our forests actively managed and healthy. In particular, the Forest Stewardship Program (FSP) of the USDA Forest Service is the gateway through which private landowners are encouraged and educated on how to burn and manage their acres to keep them healthy. Supporting FSP and other critical State and Private Forestry programs is essential to keeping up the pace of active management and prescribed burning.

2. Can you provide your perspective on whether more coordination among federal and state authorities is needed to make a meaningful difference in reducing the risks of catastrophic wildfires?

Increasing coordination between public agencies and authorities is crucial, whether that be at the state, federal or local level, to address wildfire suppression and mitigation challenges. The recently released report from the USDA Forest Service *"Toward Shared Stewardship Across Landscapes: An Outcome Based Investment Strategy"* presents an unprecedented opportunity to foster that collaboration among fire and forest managers. The forested landscape in our country is a patchwork of ownerships (private, state, federal, tribal, etc), and fire, insects, disease and other challenges know no boundaries. To make a meaningful difference in addressing any of these challenges we need to work together, draw from each other's strengths, and coordinate our resources to be most efficient with the tools we have.

3. Should air quality considerations play a greater role in informing decisions related to wildfire suppression and forestry management planning, and if so, how so?

The air quality impacts of fire on communities are becoming increasingly apparent as fires get larger and more people move closer to forested areas. These and other changes necessitate an intentional assessment of air quality impacts alongside any plans for forest management, prescribed fire, or wildfire suppression. Health and human impact data is increasingly showing that the serious risks to humans from wildfire are not confined to the flames on the ground, but that smoke impacts dozens or even hundreds of miles way could be the greatest human threat from fire.

As such, this should be a wake-up call to do more prescribed burning under manageable conditions as well as forest thinning and active management to reduce hazardous fuels. If we have the ability to do preventive work in our forests to reduce air quality impacts on communities from wildfire, then our management decisions need to be informed by that opportunity. At the federal level, National Environmental Policy Act (NEPA) analyses should include assessments of the benefits of proposed forestry projects on potential wildfire air quality emissions which would be of detriment to the environment and human communities, and that information considered by line officers in their decision making.

Questions from the Honorable Bill Flores

1. Your testimony mentions both prescribed burning and tree thinning as important tools available to make our forests healthier and less fire prone.
   a) As a forester, could you explain how cutting trees makes the remaining forests healthier?
   b) How successful has tree thinning been in Texas and the rest of the country? Should we be doing more?

In any forest, there is a finite amount of resources (water, nutrients, etc) to be shared among the trees and associated vegetation. The trees are in competition for these resources. When some of the trees are removed, this allows the remaining trees to access more water and nutrients, grow larger, and get healthier. Healthier trees are more resistant to insects and disease and the forest more resilient to wildfire. As land managers, we aim to manage each forest stand at a density that allows the trees to be as healthy as possible. When the density gets too high, we cut the smaller trees (i.e., – forest thinning) or use prescribed burning to thin out the smaller trees and brush to reduce the stand density to an ideal level for forest health.

Additionally, reducing hazardous fuels through either prescribed burning or forest thinning makes the remaining forest more resilient to wildfire. Without the ground fuels and small trees that act as "ladder fuels" that would otherwise transport fire into the upper canopy and damage or destroy the trees, wildfires in thinned managed stands take on a healthier more historically appropriate role. These fires stay closer to the ground, produce less smoke and air quality pollutants, are easier to contain, and have many ecological and wildlife benefits. This is all made possible by proper tree cutting and/or prescribed fire.

We need to be doing more tree thinning all across this country, especially on federal lands where we see millions of acres in need of restoration. There is a need for more resources at all levels to support federal land managers, state agencies, private landowners, and other stakeholders to conduct these thinning operations that benefit all of society. A critical component to getting enough thinning done on the landscape is having forest products markets for biomass and the small diameter trees that are thinned out. Often times this is the limiting factor in how much good work can be done on the ground – if there are no markets then sound management will be limited. We need to be doing more to support domestic markets for small diameter thinned wood.

In Texas, we partner with the USDA Forest Service to deliver the Forest Inventory & Analysis program. This program collects tree and forest data from all across Texas to provide meaningful information about forest health and timber availability. Because of FIA, we were able to demonstrate that

Texas has enough trees to sustainably supply a brand-new oriented strand board (OSB-similar to plywood) mill in Corrigan, TX, as well as a brand-new sawmill in Lufkin, TX. These two mills will provide hundreds of jobs and will create new markets for Texas forest owners.

As stated in my written and oral testimony, Texas was an early adopter of the Good Neighbor Authority and continues to work with our federal partners to restore healthy forests on our Texas National Forests and Grasslands through both prescribed fire and forest thinning.

Questions from the Honorable Richard Hudson

1. In your testimony you say that "our forests are currently more fire-prone than ever." Why is this the case? How did we get to the point where our forests are at such high risk of fire? Are there any regulations inhibiting the ability to lower our forests risks-of fire?

As I stated in my written testimony, our nation's forests are indeed currently more fire-prone than ever. Fire is a natural phenomenon for nearly every forest ecosystem in this country. It has shaped the occurrence and distribution of different ecosystems for centuries, simultaneously impacting the human and natural communities that live in and around those forests. However, over the past century, a culture of fire suppression has unfortunately removed the natural role of fire from the public consciousness to varying degrees in different regions. When combined with a reduced level of forest management in many areas of the country this culture has also led to the build-up of hazardous fuels and unhealthy forests to historic levels.

Of these two factors, the level of forest management is more easily addressed through policy and regulation. As discussed at the hearing, prescribed fire and active forest management are both key tools to lowering fire risk. National Environmental Policy Act (NEPA) processes and regulations have taken on an increasingly cumbersome role for federal land managers in the past three decades, straying far from the original intent of the law and delaying good and necessary forest management work that would also reduce fire risk.

Air quality regulations relative to prescribed burning can also play a role in inhibiting prescribed fire that would otherwise reduce wildfire risk. At both the federal level through the EPA, as well as at the state level, regulations need to be flexible enough to recognize that allowing smoke now in lieu of a lot of smoke later is a good policy outcome.

In: Wildfires
Editor: Dave Hawkins
ISBN: 978-1-53617-182-2
© 2020 Nova Science Publishers, Inc.

Chapter 2

PREPARE FOR FIRE SEASON[*]

United States Environmental Protection Agency

If you live in an area where the wildfire risk is high, take steps now to prepare for fire season. Being prepared for fire season is especially important for the health of children, older adults, and people with heart or lung disease.

BEFORE A WILDFIRE

- *If any family member has heart or lung disease, including asthma,* check with your doctor about what you should do during smoke events. Have a plan to manage your condition.
- *Stock up* so you don't have to go out when it's smoky. Have several days of medications on hand. Buy groceries that do not need to be refrigerated or cooked because cooking can add to indoor air pollution.

---

[*] This is an edited, reformatted and augmented version of U.S. Environmental Protection Agency Publication No. EPA- 452/F-18-001.

- *Create a "clean room"* in your home. Choose a room with no fireplace and as few windows and doors as possible, such as a bedroom. Use a portable air cleaner in the room.
- *Buy a portable air cleaner* before there is a smoke event. Make sure it has high efficiency HEPA filters and it is the right size for the room.
- *Know how you will get alerts* and health warnings, including air quality reports, public service announcements (PSAs), and social media warning you about high fire risk or an active fire.
- Ask an air conditioning professional what kind of high efficiency filters to use in your home's system and how to close the fresh-air intake if your central air system or room air conditioner has one.
- Have a supply of N95 respirators and learn how to use them. They are sold at many home improvement stores and online.
- Organize your important items ahead of time, including financial and personal documents. Know your evacuation routes and where to go if you have to evacuate. Make sure to prepare your children, and consider your pets when making an evacuation plan.

DURING A WILDFIRE

- Follow instructions from local officials to keep yourself and your family safe.
- Stay inside with the doors and windows closed. Run your air conditioner with the fresh-air intake closed ("recirculate mode") to keep smoke from getting indoors. Seek shelter elsewhere if you do not have an air conditioner and it is too warm to stay inside with the windows closed.
- Follow your health care provider's advice if you have heart or lung disease, and follow your management plan if you have one. If your symptoms worsen, reduce your exposure to smoke and contact your provider.
- Do not add to indoor air pollution. Do not burn candles or use gas, propane, wood-burning stoves, fireplaces, or aerosol sprays. Do not fry or broil meat, smoke tobacco products, or vacuum. All of these can increase air pollution indoors.
- Use a portable air cleaner to reduce indoor air pollution. Follow the manufacturer's instructions on where to put the air cleaner and when to replace the filters.
- *Reduce how much smoke you inhale.* If it looks or smells smoky outside, avoid strenuous activities such as mowing the lawn or going for a run. Wait until air quality is better before you are active outdoors.
- *Pay attention to local air quality reports and health warnings.* Smoke levels can vary a lot during the day, so you may have a chance to do errands and open up windows when air quality is better. Public service announcements give you important information such as changing conditions, cancelled events, or evacuation notices.
- *Do not rely on dust masks or bandanas* for protection from smoke. An N95 respirator can protect you if it fits snugly to your face and is worn properly. These are not recommended for children.

- *Reduce smoke in your vehicle* by closing the windows and vents and running the air conditioner in recirculate mode. Slow down when you drive in smoky conditions.

For more information:

- *Get air quality information:* Check your local news, the airnow.gov website, or your state air quality agency's website.
- *Learn about home air cleaners:* https://www.epa.gov/indoor-air-quality-iaq/guide-air-cleaners-home
- *Find certified air cleaning devices:* http://www.arb.ca.gov/research/indoor/aircleaners/certified.htm
- *Learn how to use an N95 respirator mask:* http://www.beprepared california.ca.gov/Documents/Protect%20Your%20Lungs%20Respirator.pdf
- *Learn more about wildfire smoke: How Smoke from Fires Can Affect Your Health:* https://airnow.gov/index.cfm?action=smoke.index

In: Wildfires
Editor: Dave Hawkins

ISBN: 978-1-53617-182-2
© 2020 Nova Science Publishers, Inc.

Chapter 3

REDUCE YOUR SMOKE EXPOSURE[*]

United States Environmental Protection Agency

When wildfres create smoky conditions, there are things you can do, indoors and out, to reduce your exposure to smoke. Reducing exposure is important for everyone's health — especially children, older adults, and people with heart or lung disease.

REDUCE SMOKE EXPOSURE INDOORS

- *Stay inside with the doors and windows closed.* Whether you have a central air conditioning system or a room unit, use high efciency flters to capture fne particles from smoke. Ask an air conditioning professional what type of high efciency flter your air conditioner can accept.

---

[*] This is an edited, reformatted and augmented version of U.S. Environmental Protection Agency Publication No. EPA- 452/F-18-003.

- *Seek shelter elsewhere* if you do not have an air conditioner and it is too warm to stay inside with the windows closed.
- *Do not add to indoor air pollution.* Do not burn candles or use gas, propane, wood- burning stoves, freplaces, or aerosol sprays. Do not fry or broil meat, smoke tobacco products, or vacuum. All of these can increase air pollution indoors.
- *Use a portable air cleaner to reduce indoor air pollution.* Make sure it is sized for the room and that it does not make ozone, which is a harmful air pollutant. Portable air cleaners can be used along with efcient central air systems with efcient flters to maximize the reduction of indoor particles.
- *Create a "clean room" in your home.* Choose a room with no fireplace and as few windows and doors as possible, such as a bedroom. Use a portable air cleaner in the room.
- *Have a supply of N95 respirators and learn how to use them.* They are sold at many home improvement stores and online.
- Long-term smoke events usually have periods when the air is better. When air quality improves, even temporarily, *air out your home* to reduce indoor air pollution.

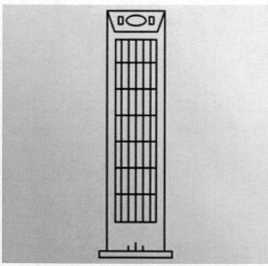

Use a portable air cleaner to reduce indoor air pollution.

REDUCE SMOKE EXPOSURE OUTDOORS

- *Take it easier during smoky times* to reduce how much smoke you inhale. If it looks or smells smoky outside, avoid strenuous activities such as mowing the lawn or going for a run.
- *Know your air quality.* Smoke levels can change a lot during the day, so wait until air quality is better before you are active outdoors. Check your state or local air quality agency's website or airnow.gov for air quality forecasts and current air quality conditions. On AirNow, you can also sign up to get email notifcations, download an air quality app, or check current fre conditions. In addition, some communities have visual range programs where you can assess smoke conditions by how far you can see.
- Have enough food and medication on hand to last several days so you don't have to go out for supplies. If you must go out, avoid the smokiest times of day.
- *Reduce smoke in your vehicle* by closing the windows and vents and running the air conditioner in recirculate mode. Slow down when you drive in smoky conditions.
- *Do not rely on dust masks or bandanas* for protection from smoke. If you must be out in smoky conditions, an N95 respirator can protect you, if it fts snugly to your face and is worn properly.
- *Have a plan to evacuate.* Know how you will get alerts and health warnings, including air quality reports and public service announcements (PSAs). Public advisories can provide important information such as changing smoke conditions and evacuation notices. Know your evacuation routes, organize your important items ahead of time, and know where to go in case you have to evacuate.

Reduce your risk of health problems

- Have enough medication and food (enough for more than 5 days) on hand.
- Follow your health care provider's advice about what to do if you have heart or lung disease.
- If you have asthma, follow your asthma management plan.
- If you feel sick, reduce your exposure to smoke and contact your health care provider.
- Pay attention to public service announcements, health advisories, and air quality advisories.

For more information

- Get air quality information: Check your local news, the airnow.gov website, or your state air quality agency's website.
- Learn about home air cleaners: https://www.epa.gov/indoor-air-quality-iaq/guide-air-cleaners-home
- Find certified air cleaning devices: http://www.arb.ca.gov/research/indoor/aircleaners/certified.htm

- Learn how to use an N95 respirator mask: http://www.beprepared california.ca.gov/ResourcesAndLinks/Languages/ Documents/English/ENG_ProtectLungsSmoke7208color.pdf
- Fires — Current Conditions: https://airnow.gov/index.cfm?action= topics.smoke_wildfires
- Learn more about wildfire smoke: Wildfire Smoke, A Guide for Public Health Officials: https://www3.epa.gov/airnow/wildfire_may 2016.pdf

In: Wildfires
Editor: Dave Hawkins

ISBN: 978-1-53617-182-2
© 2020 Nova Science Publishers, Inc.

Chapter 4

PROTECT YOURSELF FROM ASH[*]

United States Environmental Protection Agency

Protect yourself from harmful ash when you clean up after a wildfire. Cleanup work can expose you to ash and other products of the fire that may irritate your eyes, nose, or skin and cause coughing and other health effects. Ash inhaled deeply into lungs may cause asthma attacks and make it difficult to breathe.

Ash is made up of larger and tiny particles (dust, dirt, and soot). Ash deposited on surfaces both indoors and outdoors can be inhaled if it becomes airborne when you clean up. Ash from burned structures is generally more hazardous than forest ash.

AVOID ASH EXPOSURE

Avoid direct contact with ash. If you get ash on your skin, in your eyes, or in your mouth, wash it off as soon as you can.

---

[*] This is an edited, reformatted and augmented version of U.S. Environmental Protection Agency, Publication No. EPA- 452/F-18-004.

People with heart or lung disease, including asthma, older adults, children, and pregnant women should use special caution around ash.

Children and pets: Children should not be nearby while you clean up ash. Do not allow children to play in ash. Clean ash off all children's toys before use. Clean ash off pets and other animals. Keep pets away from contaminated sites.

RECOMMENDED ACTIONS

Clothing: Wear gloves, long-sleeved shirts, long pants, shoes and socks to avoid skin contact. Goggles are also a good idea Contact with wet ash can cause chemical burns or skin irritation. Change your shoes and clothing before you leave the cleanup site to avoid tracking ash offsite, into your car, or other places.

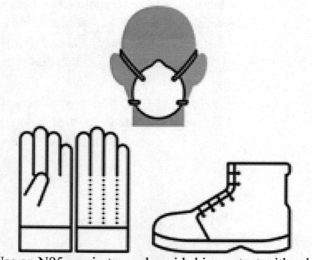

Use an N95 respirator and avoid skin contact with ash.

Protecting your lungs: Wear a tight-fitting respirator that filters ash particles from the air you breathe to help protect your lungs. Select a respirator that has been tested and approved by NIOSH and has the words

"NIOSH" and either "N95" or "P100" printed on it. These have two straps and are available online, and at many hardware stores and pharmacies. Buy respirators in a size that can be tightened over your mouth and nose with a snug seal to your face. Surgical masks and one-strap dust masks will not protect your lungs. They are not designed to seal tightly to the face. If you have heart or lung disease talk to your doctor before using a respirator or working around ash.

Cleanup: Avoid stirring up or sifting through ash as much as you can. Avoid actions that kick ash particles up into the air, such as dry sweeping. Before sweeping indoor and outdoor hard surfaces, mist them with water to keep dust down. Follow with wet mopping. Use a damp cloth or wet mop on lightly dusted areas. When you wet down ash, use as little water as you can.

Vacuum: Use a high-efficiency particulate air (HEPA)-type vacuum to clean dusty surfaces. Don't use a typical household vacuum or a shop vacuum. They will send the collected dust or ash out into the air. Don't use leaf blowers or do anything else that will put ash into the air.

Food and Water: Wash any home-grown fruits or vegetables from trees or gardens where ash has fallen. Avoid bringing food or eating at the affected site, unless you keep the food in a sealed container.

Wash your hands well before eating. Check with your drinking water provider to be sure your water is safe to drink.

Disposal: Collected ash may be disposed of in the regular trash. Ash should be stored in plastic bags or other containers to prevent it from being stirred up. If you suspect hazardous waste, including asbestos, is present, contact your local hazardous waste authorities regarding appropriate disposal. Avoid washing ash into storm drains.

For more information

- *Get air quality information:* Check your local news, the airnow.gov website, or your state air quality agency's website.
- *Learn about home air cleaners:* https://www.epa.gov/indoor-air-quality-iaq/guide-air-cleaners-home

- *Find certified air cleaning devices:* http://www.arb.ca.gov/research/indoor/aircleaners/certified.htm
- *Learn how to use an N95 respirator mask:* http://www.beprepared california.ca.gov/ResourcesAndLinks/Languages/Documents/English/ENG_ProtectLungsSmoke7208color.pdf
- *Learn more about wildfire smoke: Wildfire Smoke, A Guide for Public Health Officials:*
- https://www3.epa.gov/airnow/wildfire_may2016.pdf.

In: Wildfires
Editor: Dave Hawkins
ISBN: 978-1-53617-182-2
© 2020 Nova Science Publishers, Inc.

Chapter 5

PROTECTING CHILDREN FROM WILDFIRE SMOKE AND ASH[*]

United States Environmental Protection Agency

BACKGROUND

- Children are especially at risk for health effects from exposure to wildfire smoke and ash, mostly because their lungs are still growing.
- Wildfire concerns include the fire itself, the smoke and ash, and the chemicals from materials that have burned, such as furniture.
- Smoke can travel hundreds of miles from the source of a fire. Pay attention to local air quality reports during fire season, even if no fire is nearby.

---

[*] This is an edited, reformatted and augmented version of U.S. Environmental Protection Agency Publication No. EPA- 452/F-18-006.

HEALTH EFFECTS FROM WILDFIRE SMOKE AND ASH

- Children who breathe in wildfire smoke and ash can have chest pain and tightness; trouble breathing; wheezing; coughing; nose, throat, and eye burning; dizziness; or other symptoms.
- Children with asthma, allergies, or chronic health issues may have more trouble breathing when smoke or ash is present.

PREPARING FOR WILDFIRES

- Pay attention to local air quality reports. Stay alert to smoke-related news coverage and public health advisories.
- Look up your local Air Quality Index (AQI) on the AirNow (www.airnow.gov) web site.
- If Enviroflash is available for your area, sign up for air quality alerts. (http://www.enviroflash.info/).
- Create a "clean room" in your home. Choose a room with few windows and doors. Buy a portable air cleaner you can use in this room. *Never* use an ozone-generating air cleaner.
- Stock up on food, medicine and child care supplies before the threat of a wildfire.
- Remember that you may need to leave your home. Plan for it and prepare your children.

DURING WILDFIRES

- Continue to listen to local reports and public health warnings.
- Keep children indoors with the doors and windows closed. Use your "clean room". If you have an air conditioner, run it with the fresh-air intake *closed* to keep outdoor smoke from getting indoors. Use

- your portable air cleaner as well. Reduce health risks by avoiding strenuous activities.
- Keep the indoor air as clean as possible. Do *not* smoke. Do *not* use gas, propane, or wood- burning stoves, fireplaces, or candles. Never use ozone-generating air cleaners. *Never* use natural gas or gasoline-powered generators indoors. Do *not* use spray cans. Do *not* fry or broil meat. Do *not* vacuum. All of these can lead to poor air quality.
- A good time to open windows to air out the house and clean away dust indoors is once air quality improves (check AirNow for updates).
- Use common sense to guide your child's activity. If it looks or smells smoky outside, if local air quality is reported as poor, or if local officials are giving health warnings, wait until air quality improves before your family is active outdoors.

SPECIAL CONSIDERATIONS

- If your child has any problem breathing, is very sleepy, refuses food and water, or other health concerns, reduce his/her exposure to smoke and seek medical help right away.
- If your child has asthma, allergies, or a chronic health condition, he/she is at high risk from health effects related to wildfire smoke and ash. Seek medical advice as needed. For children with asthma, follow the asthma action plan.
- Do not rely on masks for protection from smoke. Paint, dust and surgical masks, even N95 masks, are not made to fit children and will not protect children from breathing wildfire smoke. Humidifiers or breathing through a wet washcloth do not prevent breathing in smoke.

EVACUATION

- Seek shelter in another place (e.g., public air shelter) if your family does not have an air conditioner OR air cleaner OR if it is too warm in your home to stay inside with the windows closed. Plan to take the quickest route to the shelter to limit exposure to smoke.
- Bring all medication (taken by each family member) with you.
- Reduce smoke in your vehicle by closing the windows and vents and operating the air conditioning with the fresh intake closed to keep outdoor smoke from getting into car. Never leave children in a car or truck alone.

AFTER A WILDFIRE

- Make sure ash and debris have been removed before bringing your child back to home or school.
- Children should not be doing any cleanup work. Fires may deposit large amounts of ash and dust with harmful chemicals.
- Avoid bringing polluted ash and dust back to areas used by children (such as a home or car). Remove shoes at the doorway, wash clothing separately, and change out of clothing before you have contact with your children.

For More Information

- *Learn more about wildfire smoke*: W*ildfire Smoke, A Guide for Public Health Officials*: .https://www3.epa.gov/airnow/wildfire_may 2016.pdf
- *Get air quality information:* Check the airnow.gov website, or your state air quality agency's website. *Air Quality Flag Program:* This visual tool alerts schools and organizations and their communities to the local air quality forecast. https://airnow.gov/flag.

- *Learn about home air cleaners*: https://www.epa.gov/indoor-air-quality-iaq/guide-air-cleaners-home
- *Find certified air cleaning devices:* http://www.arb.ca.gov/research/indoor/aircleaners/certified.htm Contact Poison Control at *1-800-222-1222* for emergency concerns regarding ingestion or exposure to hazards.
- *Contact your Pediatric Environmental Health Specialty Unit* with children's environmental health questions: www.pehsu.net.

Document Authored by Marissa Hauptman, MD, MPH, Laura Anderko, PhD, RN, Jason Sacks, MPH, Lora Strine, Scott Damon MAIA, Susan Stone, MS, Wayne Cascio, MD, Martha Berger, MPA. Aspects of this fact sheet were adapted from 2011 PEHSU Factsheet: Health Risks of Wildfires for Children - Acute Phase Guidance by James M. Seltzer, M.D., Mark Miller, M.D., M.P.H, and Diane Seltzer, M.A.—Region 9 Western States Pediatric Environmental Health Specialty Unit.

This document was supported in part by the American Academy of Pediatrics and the American College of Medical Toxicology and funded (in part) by the cooperative agreement award number FAIN: U61TS000237 and UG1TS000238 from the *Agency for Toxic Substances Disease Registry (ATSDR)*. The U.S. EPA supports the PEHSU by providing funds to ATSDR under Inter-Agency Agreement number DW-75-92301301. Neither U.S. EPA not ATSDR endorses the purchase of any commercial products or services mentioned in PEHSU publications.

This factsheet is dedicated in memory of Dr. James M. Seltzer as well as the first responders and others who have been affected by wildfires.

In: Wildfires
Editor: Dave Hawkins

ISBN: 978-1-53617-182-2
© 2020 Nova Science Publishers, Inc.

Chapter 6

PROTECT YOUR PETS FROM WILDFIRE SMOKE[*]

United States Environmental Protection Agency

Your pets can be affected by wildfire smoke. If you feel the effects of smoke, they probably do, too!

Smoke can irritate your pet's eyes and respiratory tract. Animals with heart or lung disease and older pets are especially at risk from smoke and should be closely watched during all periods of poor air quality.

KNOW THE SIGNS

If your animals have any of these signs, call your veterinarian:

- Coughing or gagging

---

[*] This is an edited, reformatted and augmented version of U.S. Environmental Protection Agency Publication No. EPA-452/F-19-002.

- Red or watery eyes, nasal discharge, inflammation of throat or mouth or reluctance to eat hard foods
- Trouble breathing, including open-mouth breathing, more noise when breathing, or fast breathing
- Fatigue or weakness, disorientation, uneven gait, stumbling
- Reduced appetite or thirst

RECOMMENDED ACTIONS

Even if the fire danger is not imminent, high levels of smoke may force you to stay indoors for a long time or even to evacuate. Reduce your pet's exposure to smoke as you would reduce your own.

Before the Fire Season

- Whether you have a central air conditioning system or a room unit, buy high efficiency filters you can use to capture fine particles from smoke.
- Think about creating a clean room in your house with a portable air cleaner.

When Smoke Is Present

- *Keep pets indoors* as much as you can, with doors and windows closed. Bring outdoor pets into a room with good ventilation, like a utility room, garage, or bathroom. Move potentially dangerous products, such as pesticides, out of the reach of pets.
- *Smoke is especially tough on your pet birds.* Keep them inside when smoke is present.
- *Keep indoor air clean:* do not fry or broil foods, vacuum, burn candles, use a fireplace or woodstove, or smoke tobacco products. These activities add particles to your home.
- *Spend less time outdoors and limit physical activities when it is smoky.* For example, when it's smoky, it's not a good time for you and your pet to go for a run. Let dogs and cats outside only for brief bathroom breaks if air quality alerts are in effect.

Be ready to evacuate: Include your pets in your planning. Have each pet permanently identified, for example with a microchip. Know where they will be allowed to go if there is an evacuation – not all emergency shelters accept pets. Know where your pets might hide when stressed, so you won't have to spend time looking for them in an emergency. Get pets used to their carriers and have your family practice evacuating with your pets. Covering carriers with a sheet during transport may calm a nervous pet.

If you must leave your pets behind, NEVER tie them up.

Evacuation Kit

Prepare a pet Evacuation Kit. Assemble the kit well before any emergency and store it in an easy-to-carry, waterproof container close to an exit.

- Food, water, and medicine for 7 to 10 days
- Sanitation and first aid supplies
- Important documents, such as: identification papers including proof of ownership; vaccination records; medical records and medication instructions; emergency contact list, including veterinarian and pharmacy; and a photo of your pet (preferably with you)
- Travel supplies, such as: crate or pet carrier labeled with your contact information; extra collar/harness with ID tags and leash; flashlight, extra batteries; and muzzle
- Comfort items, such as favorite toys and treats, and an extra blanket or familiar bedding

---

For more information

- Get air quality information: Check your local news, the airnow.gov website, or your state air quality agency's website.
- Reduce Your Smoke Exposure fact sheet: https://www3.epa.gov/airnow/smoke_fires/reduce-your-smoke- exposure.pdf

- Learn more about wildfire smoke: *Wildfire Smoke Guide for Public Health Officials*: https://airnow.gov/
- American Veterinary Medical Association. Get more tips and information on caring for pets and livestock during a wildfire: https://www.avma.org/public/EmergencyCare/Pages/Wildfire-Smoke-and-Animals.aspx

In: Wildfires
Editor: Dave Hawkins

ISBN: 978-1-53617-182-2
© 2020 Nova Science Publishers, Inc.

Chapter 7

PROTECT YOUR LARGE ANIMALS AND LIVESTOCK FROM WILDFIRE SMOKE[*]

United States Environmental Protection Agency

Your animals can be affected by wildfire smoke. If you feel the effects of smoke, they probably do too! High levels of smoke are harmful. Long exposure to lower levels of smoke can also irritate animals' eyes and respiratory tract and make it hard for them to breathe. Reduce your animals' exposure to smoke the same way you reduce your own: spend less time in smoky areas and limit physical activity. Animals with heart or lung disease and older animals are especially at risk from smoke and should be closely watched during all periods of poor air quality. Take the following actions to protect your large animals and livestock against wildfire smoke.

---

[*] This is an edited, reformatted and augmented version of U.S. Environmental Protection Agency Publication No. EPA-452/F-19-001.

PROTECT YOUR ANIMALS DURING SMOKE EPISODES

- Limit strenuous activities that increase the amount of smoke breathed into the lungs.
- Provide plenty of fresh water near feeding areas.
- Limit dust exposure by feeding low-dust or dust-free feeds and sprinkling or misting the livestock holding areas.
- Consider moving outdoor birds to a less smoky environment, such as a garage or basement.
- Give your livestock 4 to 6 weeks to recover fully from smoky conditions before resuming strenuous activity.
- Protect yourself, too! Think about wearing an N95 or P100 respirator while taking care of your animals.

PREPARE BEFORE A WILDFIRE

Know where to take your livestock if smoke persists or becomes severe, or if you need to evacuate. Good barn and field maintenance can reduce fire danger for horses and other livestock.

Record Keeping

- Make sure your animals have permanent identification (ear tags, tattoos, electronic microchips, brands, etc.).
- Keep pictures of animals, especially high-value animals, such as horses, up-to-date.
- Keep a list of the species, number and locations of your animals with your evacuation supplies.
- Note animals' favorite hiding spots. This will save precious rescue time!
- Keep vaccination records, medical records and registration papers with your **Evacuation Kit**.

Preparing for Evacuations

- Assemble an Evacuation Kit.
- Know where you can temporarily shelter your livestock. Contact your local fairgrounds, stockyards, equestrian centers, etc. about their policies.
- Identify trailer resources and train all livestock to load in those trailers.
- Make an evacuation plan for your animals. Plan several different evacuation routes.
- Check with local emergency management officials before you need to evacuate.

- Do not wait until the last minute. Corral animals to prepare for off-site movement.
- Above all, ensure the safety of you and your family.

Evacuation Kit

This list has just some recommended items for large animals and livestock. Your animals may have their own special requirements.

- Supply of feed, supplements and water for 7 to 10 days.
- Blankets, halters, leads, water buckets, manure fork and trash barrel.
- Copies of vaccination records, medical records and proof of ownership.
- Tools: flashlight, heavy leather gloves, rope, shovel, knife and wire cutters.
- Animal care instructions for diet and medications (for animals left at a shelter).
- Emergency cash, emergency contact list and first aid kit.

IF YOU MUST LEAVE YOUR ANIMALS BEHIND

- If evacuation cannot be accomplished in a safe and timely
- way, have a preselected, cleared area where your animals can move about.
- Open gates, cut fences, or herd livestock into areas of lower fire risk.
- Let neighbors and first responders know to be on the lookout for your animals.
- Leave enough food and water for 48 to 72 hours. Do not rely on automatic watering systems.
- Once you leave your property, do not return until told to do so by officials.

For more information

- Get air quality information: Check your local news, the airnow.gov website, or your state air quality agency's website.
- Reduce Your Smoke Exposure Fact Sheet: https://www3.epa.gov/airnow/smoke_fires/reduce-your-smoke-exposure.pdf
- Learn more about wildfire smoke: Wildfire Smoke, A Guide for Public Health Officials: https://www3.epa.gov/airnow/wildfire_may2016.pdf
- American Veterinary Medical Association. Get more tips and information on caring for pets and livestock during a wildfire:
- https://www.avma.org/public/EmergencyCare/Pages/Wildfire-Smoke-and-Animals.aspx
  https://www.avma.org/public/EmergencyCare/Pages/Large-Animals-and-Livestock-in- Disasters.aspx

INDEX

A

abatement, 251
access, 54, 242, 300, 332
accountability, 12, 16, 179
accounting, 27, 49, 56, 79, 196, 202, 326
adaptation, 54, 295, 300, 301
adults, vii, 314, 335, 339, 346
advancements, 44, 318
agencies, 11, 14, 16, 17, 18, 20, 23, 24, 25, 26, 29, 30, 31, 32, 34, 36, 37, 38, 40, 41, 42, 45, 47, 54, 57, 59, 60, 61, 62, 66, 68, 69, 81, 99, 108, 111, 133, 134, 135, 136, 140, 145, 148, 151, 155, 156, 157, 158, 159, 160, 161, 163, 164, 166, 170, 171, 172, 173, 174, 175, 176, 178, 179, 180, 181, 183, 186, 187, 292, 299, 300, 303, 305, 311, 312, 314, 318, 319, 330, 332
agency actions, 28, 314
agency collaboration, 172
agriculture, 107
air pollutants, 60, 287
air quality reports, viii, 336, 337, 341, 349, 350
air quality risks, vii, 3, 4, 5, 6, 151

air temperature, 249
Alaska, 52, 105, 206, 218, 231, 264, 280, 286
algorithm, 251
alters, 109, 114
ambient air, 49, 232, 250, 312
American Bar Association, 188
Americans with Disabilities Act, 131
ammonium, 219, 229
appetite, 356
appropriations, 8, 54, 55, 72, 299, 318
arbitration, 18
asbestos, 347
assessment, 74, 211, 331
assets, 17, 84
asthma, viii, 3, 5, 19, 22, 41, 53, 64, 78, 151, 263, 264, 315, 335, 342, 345, 346, 350, 351
asthma attacks, viii, 345
atmosphere, 25, 31, 80, 98, 175, 241, 260, 282, 297
at-risk populations, 56, 325
attitudes, 117
attribution, 191, 194, 197, 201, 205, 208

authorities, 34, 36, 37, 156, 157, 160, 164, 179, 318, 319, 324, 329, 330, 347
authority, 38, 60, 68, 84, 99, 158, 325, 330
aversion, 103, 104, 139, 167, 183, 186
awareness, 66, 219, 293, 294, 306, 329

B

backlash, 149
banks, 277
barriers, 16, 56, 78, 89, 103, 104, 132, 133, 134, 136, 139, 140, 148, 160, 161, 163, 168, 171, 174, 181, 182, 183, 184, 185, 186, 187, 188
base, 114, 127, 295
batteries, 358
bedding, 358
beetles, 280, 285, 286, 287
behaviors, 139
beneficial effect, 116
benefits, 16, 18, 42, 74, 85, 86, 91, 98, 99, 103, 108, 110, 111, 115, 118, 119, 120, 172, 189, 263, 275, 297, 298, 300, 331, 332
benzene, 219
bias, 166, 204
biodiversity, 46, 56
biomass, 104, 111, 124, 126, 201, 224, 225, 252, 332
birds, 114, 357, 362
bison, 276
blame, 80, 255, 260
boreal forest, 105, 108, 121, 206
bounds, 194, 205
breathing, 51, 86, 350, 351, 356
breeding, 114, 276, 287
budget line, 136, 179
Bureau of Indian Affairs, 164
Bureau of Land Management, 44, 108, 131, 136

burn, 25, 26, 31, 36, 39, 40, 46, 47, 50, 62, 64, 66, 68, 70, 74, 75, 79, 80, 85, 86, 88, 93, 95, 98, 99, 104, 108, 111, 112, 115, 116, 117, 119, 120, 128, 132, 133, 134, 136, 145, 146, 147, 149, 150, 152, 155, 156, 157, 158, 159, 161, 162, 163, 164, 166, 167,168, 169, 170, 172, 173, 174, 175, 177, 178, 180, 181, 182, 183, 184, 185, 186, 187, 212, 224, 227, 242, 274, 277, 280, 283, 289, 292, 295, 300, 311, 313, 315, 329, 330, 337, 340, 357
burnout, 278
businesses, 15, 297

C

cambium, 123
campaigns, 244
carbon, 53, 55, 57, 74, 77, 84, 143, 193, 202, 208, 217, 219, 224, 225, 227, 247, 248, 260, 264, 273, 282, 297, 300, 302, 303, 324
carbon dioxide, 217, 273, 282
carbon emissions, 53, 77, 84, 193, 217, 224, 225, 260
carbon monoxide, 53, 74, 143
cardiovascular morbidity, 223
case studies, 136, 140
challenges, 4, 6, 9, 18, 46, 69, 85, 88, 91, 95, 99, 103, 112, 116, 132, 133, 134, 138, 140, 150, 164, 169, 179, 180, 181, 182, 183, 184, 185, 186, 187, 188, 189, 297, 309, 314, 330
changing environment, 305
checks and balances, 169
children, v, vii, viii, 21, 23, 41, 53, 64, 309, 314, 335, 336, 337, 339, 346, 349, 350, 351, 352, 353
circulation, 105, 106
cities, 63, 255, 276

Index

citizens, 12, 15, 16, 20, 24, 25, 26, 27, 29, 35, 37, 42, 47, 61, 77, 329
clean air, 52, 91, 244
Clean Air Act, 7, 33, 50, 56, 95, 116, 132, 133, 140, 142, 143, 153, 264
clean energy, 303
cleaning, 338, 342, 348, 353
cleanup work, vii, 345, 352
climate change issues, 84
climates, 290
clothing, 346, 352
collaboration, 30, 49, 56, 57, 171, 180, 297, 311, 312, 313, 314, 323, 330
College Station, 327
combustion, 41, 64
commercial, 260, 269, 324, 325, 353
common sense, 79, 302, 351
communication, 67, 86, 99, 135, 141, 149, 154, 172, 173, 174, 177, 180, 181, 182, 183, 184, 185, 187, 309, 311, 313
communication strategies, 174
communities, vii, 3, 5, 10, 12, 15, 18, 19, 22, 25, 29, 34, 37, 40, 42, 44, 45, 46, 47, 51, 52, 56, 57, 59, 60, 61, 62, 63, 64, 65, 67, 68, 78, 85, 86, 96, 100, 101, 109, 111, 117, 125, 126, 128, 138, 148, 152, 159, 181, 210, 259, 272, 273, 274, 275, 276, 283, 292, 294, 296, 297, 300, 301, 306, 309, 312, 314, 318, 320, 323, 326, 329, 331, 333, 341, 352
compilation, 108
complement, 201
complexity, 104, 115, 218, 241, 249
compliance, 3, 5, 17, 36, 47, 95, 131, 142, 144, 158
composition, 103, 109, 122, 127, 242, 247, 296
computing, 211
conditioning, 336, 339, 352, 357
conflagration, 265
conflict, 154, 170, 193
conflict of interest, 193

conifer, 110, 122, 123, 124, 126, 295
consciousness, 40, 63, 95, 333
consent, 102
conservation, 52, 108, 120, 126, 134, 170, 270, 301
conserving, 120
Consolidated Appropriations Act, 17
constituents, 4, 6, 10, 12, 95
conversations, 49, 73, 87, 121
cooking, 256, 315, 335
cooperation, 158, 173, 318
cooperative agreements, 174
coordination, 29, 46, 134, 145, 156, 182, 304, 305, 306, 309, 311, 312, 313, 317, 319, 322, 323, 325, 328, 330
correlation, 140, 191, 205
correlations, 196, 205
cost, 17, 18, 45, 97, 174, 257, 261, 283, 286, 299, 300, 318
coughing, viii, 41, 64, 314, 345, 350, 355
covering, 7, 52, 55, 203, 275
crown fires, 110, 291
cultivation, 107
culture, 40, 63, 139, 166, 178, 333
current limit, 120

D

danger, 147, 191, 192, 198, 199, 203, 207, 210, 211, 276, 280, 357, 363
data set, 262
database, 46, 246
death rate, 224
deaths, 213, 214, 224, 238, 239, 240, 243
decision-making process, 54
decomposition, 208
deficit, 49, 112, 190, 191, 202, 206, 207, 323
deforestation, 80, 225, 227, 243
degradation, 214, 215, 233, 234, 243
demographic change, 219

Department of Agriculture, 67, 127, 189, 277, 294, 296, 299
Department of Energy, 245
destruction, 274
detectable, 196, 202
developing countries, 219
directives, 143, 305
directors, 37, 59, 61, 140
dirt, viii, 75, 278, 345
disaster, 17, 34, 54, 256, 258, 261, 272, 299, 300, 323
disaster assistance, 34, 54
disaster relief, 256
dispersion, 98, 136, 149, 152, 156, 177
displacement, 117
distribution, 40, 63, 194, 207, 217, 234, 313, 333
District of Columbia, 37, 61
diversity, 109, 123, 125, 131, 132, 140, 277
drought, 19, 24, 27, 34, 36, 44, 70, 169, 192, 201, 208, 209, 210, 212, 256, 272, 275, 280, 286, 287, 289, 290, 299
dust, viii, 219, 287, 337, 341, 345, 347, 351, 352, 362

E

ecological processes, 112, 146
ecological restoration, 120, 293
ecology, 116, 121, 123, 125, 127, 128
economic consequences, 73
economic growth, 219
ecosystem, 36, 37, 40, 59, 61, 64, 67, 75, 84, 85, 103, 105, 109, 111, 120, 124, 125, 126, 128, 129, 144, 214, 291
ecosystem restoration, 103, 111
education, 41, 60, 64, 101, 103, 118, 184, 219, 271
educational materials, 66
e-mail, 304, 308, 316, 321, 327

emergency, 8, 16, 25, 29, 34, 54, 78, 287, 353, 358, 363, 364
emergency management, 363
emission, 26, 155, 158, 217, 218, 225, 227, 229, 233, 234, 241, 244
employees, 66, 135, 162, 166, 183, 184
endangered, 109, 117, 122, 170, 276
endangered species, 117, 170
energy, 34, 74, 92, 192, 217, 219, 249, 265
energy efficiency, 92
environment, 75, 90, 138, 193, 256, 331, 362
environmental change, 210
environmental degradation, 219
environmental impact, 86, 97
environmental movement, 81
environmental organizations, 182
Environmental Protection Agency, v, vii, 15, 29, 42, 65, 143, 214, 253, 287, 320, 335, 339, 345, 349, 355, 361
environmental quality, 60, 69, 86, 141, 312
environments, 117, 191, 194
epidemic, 285
epidemiology, 241
equilibrium, 217
equipment, 59, 88, 108, 136, 170, 177, 185, 187
evacuation, 52, 100, 283, 336, 337, 341, 358, 363, 364
evaporation, 255
evapotranspiration, 192, 208, 211, 280, 290
evidence, 19, 22, 41, 103, 251, 276
evolution, 209
exclusion, 9, 36, 40, 55, 95, 104, 109, 110, 114, 117, 119, 122, 124, 198, 200
executive branch, 139
expertise, 61, 65, 66, 84, 89, 166, 178, 314
exposure, vii, viii, 19, 22, 24, 28, 42, 213, 215, 216, 223, 224, 232, 237, 238, 239, 241, 242, 243, 244, 245, 246, 248, 250, 337, 339, 342, 349, 351, 352, 353, 357, 358, 361, 362, 365

extinction, 220, 221, 249, 251
eyes, viii, 16, 314, 345, 355, 356, 361

F

fear, 52, 119, 163
federal agency, 135, 139
federal government, 38, 57, 179, 256, 260, 296
Federal Government, 59, 81, 88
federal law, 135
filters, 25, 29, 53, 99, 336, 337, 346, 357
financial, 37, 59, 61, 302, 336
financial incentives, 302
fire season, vii, viii, 3, 5, 8, 10, 12, 13, 18, 20, 22, 27, 34, 35, 38, 39, 43, 50, 59, 62, 63, 90, 91, 97, 106, 114, 116, 123, 124, 158, 169, 172, 182, 197, 207, 272, 276, 280, 286, 299, 305, 306, 307, 335, 349
fire suppression, 11, 14, 24, 27, 40, 53, 63, 64, 103, 108, 116, 120, 161, 184, 190, 191, 193, 202, 272, 274, 277, 292, 301, 305, 333
first aid, 358, 364
first responders, 353, 364
fiscal year, 8, 69, 72, 164, 323
fish, 174, 273, 283, 287
fishing, 15
flame, 118
flammability, 103, 115, 191, 203
flexibility, 30, 31, 57, 133, 136, 172, 175, 180, 324
food, 106, 107, 279, 341, 342, 347, 350, 351, 364
force, 81, 84, 120, 154, 274, 357
Ford, 123, 213, 222, 240, 241, 246
forest ash, viii, 345
forest ecosystem, 40, 63, 272, 324, 333
forest fire, 12, 13, 33, 34, 41, 64, 69, 75, 92, 100, 125, 127, 190, 191, 192, 193, 194, 196, 199, 200, 201, 202, 205, 207, 208, 211, 256, 273, 287, 295, 323
forest management, 7, 8, 12, 15, 16, 17, 18, 20, 21, 22, 23, 33, 35, 38, 40, 52, 53, 55, 58, 60, 62, 63, 70, 71, 72, 74, 75, 83, 89, 95, 101, 108, 116, 121, 188, 258, 265, 272, 274, 292, 295, 297, 301, 304, 305, 306, 308, 310, 311, 317, 319, 322, 323, 328, 329, 330, 331, 333
forest resources, 301
forest restoration, 45, 54, 55, 114, 180, 272, 273, 274, 290, 292, 294, 295, 296, 299, 300, 301, 318, 323, 324, 325, 326
formula, 212
foundations, 124
fruits, 347
fuel consumption, 114, 123
fuel distribution, 112
fuel efficiency, 33
fuel loads, 20, 23, 56, 70, 166, 169, 191, 292, 295, 318, 320
fuel management, 103, 172
funding, 8, 16, 25, 29, 32, 42, 45, 48, 49, 50, 51, 53, 54, 55, 56, 57, 58, 69, 72, 78, 79, 81, 82, 83, 85, 88, 92, 94, 101, 104, 117, 120, 130, 132, 133, 135, 136, 146, 160, 161, 162, 164, 165, 175, 176, 178, 181, 183, 184, 185, 186, 205, 258, 260, 261, 273, 298, 299, 300, 301, 318, 323, 324
funds, 17, 34, 78, 79, 164, 173, 175, 186, 272, 299, 301, 323, 353

G

gait, 356
geography, 60, 67, 204
global scale, 248
global warming, 33, 209, 256, 257
goal-setting, 141
God, 90
Gore, Al, 80

governments, 28, 101, 306
governor, 99
grasses, 125, 275
grasslands, 57, 128, 278, 297
greenhouse, 57, 217, 260, 274, 302
greenhouse gas, 57, 217, 260, 274, 302
greenhouse gas emissions, 57
greenhouse gases, 260, 274
grid resolution, 251
gross domestic product, 215
growth, 34, 98, 122, 123, 203, 217, 219, 221, 275
guidance, 50, 56, 73, 81, 139, 178, 185, 187, 326
guiding principles, 312

H

habitat, 53, 55, 67, 70, 77, 109, 110, 114, 117, 170, 174, 175, 187, 259, 276, 278, 283, 287, 295, 296, 297, 324
habitats, 273, 274, 277, 278
hardwood forest, 105, 117, 125, 127, 289
harmful ash, vii, 345
harvesting, 33, 70, 107, 295
Hawaii, 218, 231
hazardous waste, 347
hazards, 82, 353
haze, 140, 142, 144, 221, 234, 235, 287
headache, 50
health, vii, viii, 3, 5, 7, 9, 12, 13, 16, 17, 18, 20, 21, 23, 24, 25, 27, 29, 30, 31, 32, 37, 41, 42, 44, 45, 47, 48, 52, 53, 56, 57, 58, 61, 64, 67, 72, 77, 100, 144, 151, 154, 169, 173, 210, 213, 214, 216, 219, 222, 228, 237, 238, 240, 241, 242, 244, 250, 252, 260, 274, 278, 288, 293, 296, 297, 298, 301, 306, 307, 309, 312, 313, 314, 318, 323, 324, 325, 329, 330, 332, 335, 336, 337, 339, 341, 342, 345, 349, 350, 351, 353
health care, 16, 242, 337, 342
health care system, 16
health effects, viii, 41, 64, 216, 219, 222, 228, 237, 240, 345, 349, 351
health impacts, vii, 3, 5, 24, 25, 27, 29, 41, 56, 151, 214, 242, 252, 309, 314, 325
health problems, 342
health risks, 42, 307, 351
heart disease, 41, 64, 264
herpetofauna, 109
heterogeneity, 113, 115
history, 7, 9, 11, 44, 60, 65, 88, 103, 114, 123, 126, 166, 189, 209, 255, 276, 278, 286
Holocene, 121, 126
homeowners, 86
homes, 9, 21, 23, 30, 33, 39, 62, 79, 92, 101, 119, 123, 184, 274, 283, 299, 309, 313
horses, 363
House, vii, 1, 12, 17, 18, 19, 26, 37, 51, 61, 73, 74, 76, 261, 303, 304, 307, 308, 310, 315, 316, 320, 321, 322, 326, 327
House of Representatives, vii, 1, 17, 51, 303, 307, 315, 320, 326
House Report, 17
housing, 108, 274, 277, 301
human, 7, 8, 11, 14, 40, 41, 52, 63, 64, 71, 91, 96, 98, 105, 106, 111, 149, 150, 151, 154, 176, 184, 190, 191, 193, 202, 208, 209, 214, 215, 245, 246, 248, 255, 256, 260, 261, 263, 264, 277, 290, 296, 331, 333
human activity, 191, 208, 260, 296
human health, 7, 41, 63, 64, 149, 150, 151, 154, 176, 202, 261, 264
human interactions, 215
humidity, 78, 98, 119, 204, 215, 221, 281
hurricanes, 8, 51, 54, 92, 100, 272, 299

I

ideal, 154, 170, 256, 332
identification, 358, 363
ignitability, 191
ignition source, 105, 107
image, 271
immigration, 107
impact assessment, 222, 250
improvements, 72, 213, 214, 237, 243
independence, 269
indirect effect, 285
individuals, 41, 64, 65, 92, 151, 161, 179, 261, 309, 314
industrial emissions, 229
industrialized countries, 219
industry, 14, 44, 45, 69, 92, 116, 149, 182, 260, 294, 297
infancy, 100
infestations, 18, 55, 285, 286
inflammation, 356
infrastructure, 34, 108, 111
ingestion, 353
ingredients, 68
insane, 259
insects, 17, 18, 85, 99, 265, 285, 330, 332
integration, 42, 103
intelligence, 65, 275
interagency coordination, 177
interagency relationships, 155
interest groups, 20, 22
interface, 79, 96, 97, 100, 151, 277
intrusions, 148, 183, 185
investment, 40, 90, 133, 177, 186, 305
investments, 44, 45, 59, 302, 318
irrigation, 259
issues, 7, 8, 19, 20, 22, 23, 25, 29, 32, 42, 44, 45, 46, 49, 50, 65, 66, 67, 74, 76, 82, 83, 86, 95, 98, 101, 103, 139, 150, 154, 157, 158, 162, 174, 184, 210, 265, 297, 318, 320, 350

J

Jordan, 2
journalism, 257
jurisdiction, 3, 5, 62, 77, 157, 323
justification, 66
juveniles, 114

K

kicks, 51
kill, 88, 113, 154, 287

L

lakes, 255, 259
landfills, 34
landscape, 39, 44, 54, 56, 68, 72, 76, 78, 97, 105, 107, 109, 111, 116, 118, 120, 122, 127, 128, 134, 152, 156, 161, 162, 163, 165, 166, 168, 169, 170, 172, 181, 182, 185, 189, 191, 210, 216, 228, 247, 282, 319, 326, 330, 332
Late Pleistocene, 122
laws, 8, 41, 64, 73, 116, 117, 128, 132, 139, 140, 142, 168, 255, 258
laws and regulations, 64
lawyers, 164
lead, 64, 109, 112, 113, 114, 116, 119, 140, 143, 149, 170, 219, 229, 233, 241, 243, 244, 256, 264, 278, 282, 325, 351
leadership, 48, 133, 135, 136, 139, 166, 176, 273
legal protection, 168
legislation, 18, 76, 118, 154
lesson plan, 271
life cycle, 295
life expectancy, 246
light, 106, 140, 169, 221, 251
limited liability, 168

litigation, 18, 20, 22
livestock, v, viii, 297, 359, 361, 362, 363, 364, 365
loans, 261
local community, 29
local government, 45, 60, 66, 79, 81, 92, 100, 101, 174, 319
logging, 11, 16, 48, 80, 107, 108, 260, 295
longleaf pine, 113, 122, 125, 128, 290
love, 51, 74, 97
lung disease, vii, viii, 315, 335, 337, 339, 342, 346, 347, 355, 361
lung function, 264

M

magnitude, 120, 234, 326
majority, 28, 67, 68, 91, 111, 114, 227
management committee, 60
manipulation, 120
manufacturing, 149, 182, 263
manure, 364
mapping, 126, 210
mass, 220, 221, 222, 249, 264
materials, viii, 66, 107, 292, 296, 301, 349
matrix, 67
measurement, 136, 177
medical, 41, 64, 351, 358, 363, 364
medication, 315, 341, 342, 352, 358
medicine, 107, 350, 358
Mediterranean, 100, 278
Mediterranean climate, 100
Mexico, 109, 111, 119, 125, 127, 140, 145, 147, 157, 185, 218, 224, 232, 243, 278
migration, 219
missions, 38, 44, 62, 312
models, 50, 175, 177, 193, 201, 204, 208, 210, 215, 225, 241, 242, 243, 250, 297, 301
Moderate Resolution Imaging Spectroradiometer, 192

Moderate Resolution Imaging Spectroradiometer (MODIS), 192
moisture, 39, 94, 104, 114, 115, 119, 201, 203, 209, 211, 215, 227, 272
mole, 71
Montana, 4, 6, 10, 11, 13, 16, 25, 30, 31, 34, 35, 36, 37, 38, 39, 41, 43, 44, 45, 46, 47, 53, 73, 89, 92, 93, 98, 99, 100, 123, 134, 136, 140, 145, 156, 157, 172, 173, 177, 183, 184, 264, 271, 275, 276, 316, 317, 318, 320
morbidity, 223
mortality, 19, 22, 75, 117, 123, 127, 193, 201, 209, 221, 222, 223, 224, 240, 242, 244, 245, 246, 247, 248, 250, 251, 252, 280, 285, 286
mortality rate, 75, 224, 240
mountain ranges, 105
multiplier, 265

N

National Academy of Sciences, 8, 102, 189, 245, 256, 262, 276, 277, 281
National Aeronautics and Space Administration, 205
National Ambient Air Quality Standards, 56, 95, 142, 143, 253, 263, 326
National Environmental Policy Act (NEPA), 17, 331, 333
National Forest System, 45, 173, 188
National Park Service, 108, 123, 164
national parks, 93, 143, 253
national security, 219
Native Americans, 103, 107, 108, 121, 126
native population, 107
native species, 278
natural disaster, 8, 34, 175
natural disasters, 8, 34
natural gas, 351
natural resource management, 120, 126

Index

natural resources, 37, 121
negative effects, 90
nitrogen, 53, 143
nitrogen dioxide, 143
North America, 102, 103, 104, 105, 106, 107, 108, 110, 113, 117, 119, 124, 126, 189, 206, 208, 209, 211, 224, 231
nose, viii, 41, 64, 345, 347, 350

O

obstacles, 56, 272
obstruction, 18
oceans, 80
officials, 4, 6, 46, 67, 309, 314, 337, 351, 363, 364
oil, 34, 185, 302
older pets, viii, 355
openness, 117
operations, 21, 34, 115, 127, 332
opportunities, 4, 6, 36, 41, 44, 48, 54, 68, 132, 134, 135, 136, 137, 139, 140, 150, 161, 166, 170, 171, 172, 173, 174, 175, 176, 178, 180, 181, 182, 183, 184, 185, 186, 187, 188, 297, 311, 318, 319
organize, 100, 341
outreach, 86, 174, 179, 182, 184, 185, 187, 313
oversight, 12, 16, 51, 53, 58, 181, 182, 183, 184, 185, 186, 187, 188, 324, 325
ownership, 44, 67, 72, 117, 170, 183, 358, 364
ozone, 53, 143, 182, 183, 245, 246, 248, 251, 252, 315, 340, 350, 351

P

pain, 314, 350
parallel, 306
participants, 28, 136, 312
pathway, 217, 219
pathways, 139, 178, 201, 212, 247
peat, 216
percentile, 199, 203, 211
permission, iv, 152, 265
permit, 47, 67, 145, 146, 152, 154, 156, 168, 304, 308, 316, 321, 327
personal relationship, 173
personality, 168
pests, 99, 280
pets, v, viii, 336, 346, 355, 357, 358, 359, 365
pH, 283
physical activity, viii, 361
pitch, 125, 286
pitch pine, 125, 286
plant growth, 256
plants, 34, 96, 107, 123, 255, 260, 278
PM, 41, 42, 53, 56, 64, 65, 155, 157, 181, 187, 213, 215, 221, 228, 234, 236, 242, 243, 244, 320, 325
pneumonia, 53
policy, 34, 103, 111, 120, 132, 135, 136, 138, 139, 140, 142, 154, 174, 176, 182, 185, 251, 269, 273, 307, 324, 326, 333, 334
policy issues, 182
policymakers, 4, 79, 90, 136
political leaders, 183
politics, 90
pollutant, 340
pollutants, 143, 332
polluters, 33
pollution, 7, 9, 19, 22, 24, 27, 34, 47, 53, 56, 69, 78, 96, 116, 132, 143, 145, 148, 149, 184, 214, 247, 248, 249, 250, 251, 262, 263, 264, 265, 273, 287, 300, 303, 315, 326, 335, 337, 340
ponds, 259
poor air quality, viii, 33, 116, 132, 148, 351, 355, 361
population, 14, 25, 30, 107, 132, 148, 149, 157, 182, 209, 213, 215, 216, 219, 220,

222, 223, 224, 227, 228, 231, 236, 237, 238, 239, 242, 243, 244, 247, 248, 276, 284, 287
precipitation, 53, 57, 106, 156, 193, 195, 199, 203, 209, 210, 211, 215, 227, 281, 290, 296
premature death, 19, 22, 33, 213, 214, 222, 223, 224, 237, 238, 239, 241, 242, 243, 264, 287
preparedness, 161
prescribed burning, 3, 5, 25, 30, 31, 32, 36, 42, 46, 60, 61, 62, 66, 67, 68, 82, 84, 85, 87, 94, 95, 99, 103, 110, 111, 114, 115, 116, 117, 119, 120, 122, 123, 128, 144, 147, 152, 155, 158, 163, 166, 167, 169, 172, 177, 178, 188, 304, 305, 308, 310, 311, 317, 322, 328, 329, 330, 331, 332, 334
preservation, 46, 128
president, 4, 48, 52, 59, 255, 257, 258, 259, 260
presidential campaign, 259
prevention, 78, 85, 95, 108, 306
principles, 25, 30, 121, 301, 305, 311
professionals, 60, 71
profit, 175
program staff, 41
project, 17, 18, 55, 97, 131, 134, 136, 139, 142, 143, 159, 167, 170, 201, 212, 243, 244, 247, 250, 281, 301
propaganda, 108
protection, 38, 59, 61, 62, 99, 128, 143, 149, 170, 174, 187, 313, 337, 341, 351
public awareness, 120
public domain, 130
public health, 3, 5, 16, 20, 23, 25, 28, 30, 31, 41, 46, 51, 56, 57, 73, 77, 92, 151, 154, 175, 186, 244, 263, 287, 302, 309, 311, 312, 313, 314, 326, 350
public interest, 330
public opinion, 154
public safety, 175
public sector, 188
public service, 336, 341, 342
pulmonary diseases, 3, 5

Q

qualifications, 179, 186
qualitative differences, 240
quality standards, 36, 43, 47, 49, 312, 320
quantification, 201

R

racial minorities, 263
radiation, 193, 203
rainfall, 34
rangeland, 30, 44, 67, 174, 318
recognition, 110, 119, 299
recommendations, iv, 26, 31, 44, 48, 53, 57, 73, 136, 141, 156, 157, 251, 313, 314, 318
recovery, 42, 106, 227, 258, 261, 277, 282, 296
recreation, 41, 63, 64, 272, 287, 297, 299
redundancy, 76
regression, 194, 196, 200, 205, 211
regrowth, 282
regulations, 4, 6, 60, 68, 139, 143, 144, 150, 156, 170, 171, 180, 253, 264, 265, 325, 329, 333, 334
regulatory agencies, 32, 36, 47, 67, 141, 173
regulatory bodies, 135
relief, 12, 16, 19, 54
renewable energy, 34, 303
requirements, 17, 133, 134, 142, 143, 144, 154, 157, 159, 163, 170, 175, 178, 179, 187, 211, 364
researchers, 32, 34, 63, 68, 256, 263, 264, 265, 281
resilience, 44, 45, 53, 55, 57, 115, 272, 273, 293, 295, 298, 301, 318, 323, 324, 325

resolution, 204, 216, 224, 241, 252, 254
resources, 14, 16, 18, 24, 29, 38, 54, 59, 60, 61, 62, 66, 68, 70, 71, 92, 94, 96, 97, 119, 127, 133, 134, 135, 136, 146, 154, 161, 162, 163, 164, 165, 174, 178, 179, 181, 183, 184, 185, 186, 187, 245, 271, 272, 279, 299, 301, 303, 313, 323, 330, 332, 363
respirator, 337, 338, 341, 343, 346, 348, 362
respiratory tract, viii, 355, 361
response, 8, 28, 32, 38, 54, 62, 72, 90, 123, 124, 221, 222, 224, 240, 250, 251, 313
restoration, 38, 44, 45, 48, 49, 50, 51, 52, 53, 54, 55, 81, 92, 101, 111, 112, 114, 115, 125, 127, 128, 174, 175, 256, 272, 273, 274, 292, 294, 295, 296, 297, 298, 300, 301, 318, 319, 323, 324, 325, 330, 332
restoration programs, 324
restrictions, 138, 157, 162, 170, 187, 260
risk assessment, 249
risks, vii, 3, 4, 5, 6, 17, 52, 53, 54, 55, 56, 67, 72, 78, 111, 117, 119, 139, 151, 152, 153, 244, 272, 274, 299, 300, 304, 306, 309, 312, 313, 317, 319, 322, 325, 326, 328, 329, 330, 331, 333
root, 114
routes, 100, 306, 336, 341, 363
rules, 87, 89, 102

S

safety, 7, 9, 21, 25, 30, 47, 98, 100, 329, 364
scaling, 274, 300
scattering, 221
school, 12, 15, 24, 28, 29, 352
school activities, 12
science, 16, 20, 22, 26, 32, 40, 53, 58, 64, 65, 70, 73, 79, 101, 103, 104, 125, 150, 210, 251, 254, 260, 265, 269, 271, 279, 280, 286, 295, 297
scientific knowledge, 103
seeding, 162
seedlings, 110, 111
sensitivity, 191, 201, 211, 218, 220, 222, 223, 224, 238, 280
shade, 109, 110
shape, 111, 121, 222, 251
shelter, 279, 337, 340, 352, 363, 364
shortage, 259
shortness of breath, 41, 64, 78, 314
showing, 110, 331
shrubland, 124, 278
significance level, 198
simulations, 208, 213, 216, 217, 218, 219, 224, 226, 227, 228, 229, 230, 232, 233, 234, 235, 236, 237, 238, 240, 241, 242, 244
sinuses, 314
SIP, 142, 144, 186
skin, viii, 345, 346
smoke and ash, vii, viii, 349
smoke exposure, 28, 42, 219, 223, 228, 244, 252
smoke plume, 241
smoky conditions, vii, 338, 339, 341, 362
social acceptance, 120
software, 42, 64
solution, 77, 83, 305, 311, 318
soot, viii, 51, 81, 287, 345
specialists, 67, 164
species, 43, 103, 109, 110, 112, 114, 115, 117, 119, 120, 122, 170, 185, 219, 221, 229, 231, 272, 278, 286, 290, 295, 296, 297, 300, 363
spending, 17, 95, 161, 260, 272, 323
staffing, 32, 164, 311
Stafford Act, 54
stakeholder groups, 174

stakeholders, 26, 32, 44, 45, 57, 73, 139, 183, 294, 301, 311, 312, 313, 314, 318, 323, 332
state authorities, 304, 306, 309, 312, 317, 319, 322, 325, 328, 330
State Implementation Plan, 142, 143
state laws, 140
state of emergency, 276
states, 5, 6, 27, 37, 38, 44, 45, 46, 47, 52, 53, 56, 57, 61, 65, 66, 67, 68, 118, 125, 131, 132, 133, 135, 136, 140, 142, 143, 144, 145, 146, 148, 151, 158, 160, 161, 163, 167, 168, 171, 172, 173, 174, 175, 177, 180, 181, 183, 184, 185, 233, 264, 275, 287, 305,309, 313, 318, 319, 323, 325, 326
statistics, 39, 76, 109, 111, 188, 205
statutes, 143, 150
statutory authority, 67
storage, 300, 302, 324
stoves, 337, 340, 351
strategic management, 103, 116
strategic planning, 136, 180, 182
stress, 113, 209, 280, 286
stressors, 48, 71
structure, 104, 109, 122, 127, 133, 144, 145, 147, 166, 184
styles, 11, 14
subtraction, 204
succession, 115, 203
suicide, 154
sulfate, 219, 221, 229, 249
sulfur, 143
sulfur dioxide, 143
supervisors, 178
suppression, 10, 11, 12, 14, 17, 40, 41, 63, 94, 101, 108, 109, 112, 120, 123, 127, 154, 162, 166, 172, 191, 201, 215, 219, 227, 299, 305, 306, 307, 309, 312, 317, 319, 322, 325, 328, 330, 331
surface area, 104
surplus, 207
surrogates, 120
susceptibility, 114, 280
sustainability, 219, 301
synthesis, 123, 124

T

tactics, 305
target, 167, 296
taxpayers, 18, 94
teams, 38, 134, 171, 176, 180, 186
technical assistance, 82, 185, 330
techniques, 8, 20, 22, 62, 83, 93, 94, 177, 185, 293, 295, 311
technological change, 217
technologies, 303
temperature, 98, 101, 106, 190, 193, 194, 202, 203, 204, 209, 210, 211, 215, 227, 281, 296
temporary housing, 261
tension, 111, 116, 117
terrestrial ecosystems, 103
thinning, 16, 20, 22, 23, 40, 62, 70, 81, 85, 114, 120, 161, 167, 184, 260, 295, 307, 320, 328, 331, 332, 333
thoughts, 73
threats, 18, 45, 54, 57, 101, 202
timber production, 111
time periods, 216, 218, 224, 235, 237, 239, 280
time series, 192, 203, 204, 205
tobacco, 337, 340, 357
tourism, 12, 15, 41, 63, 64, 148
toxicity, 223, 242
training, 13, 59, 87, 118, 152, 162, 169, 176, 178, 185
transition period, 217
transparency, 136, 175, 265
transport, 231, 237, 238, 245, 332, 358
transportation, 282

Index

treatment, 16, 18, 42, 44, 45, 67, 76, 99, 101, 127, 296
tropical rain forests, 80
turnout, 58

U

U.S. Environmental Protection Agency, vii, 214, 287, 335, 339, 345, 349, 355, 361
uniform, 112
uninsured, 261
United States, v, 3, 5, 8, 51, 72, 102, 123, 124, 126, 128, 190, 191, 193, 194, 195, 197, 199, 201, 203, 206, 207, 208, 209, 210, 211, 212, 213, 214, 216, 218, 223, 224, 225, 226, 227, 228, 229, 232, 233, 234, 236, 237, 238, 240, 241, 242, 243, 244, 245, 246, 247, 248, 249, 251, 252, 253, 254, 262, 263, 264, 276, 277, 280, 281, 285, 297, 299, 303, 307, 315, 320, 326, 335, 339, 345, 349, 355, 361
universities, 90
updating, 42, 65, 97, 169, 320
urban, 59, 63, 79, 86, 97, 100, 126, 155, 219, 237, 241, 244, 247, 277
urban areas, 63, 219
urbanization, 116, 219
USDA, 17, 57, 103, 122, 123, 124, 125, 127, 130, 188, 189, 251, 318, 319, 330, 332

V

vacuum, 315, 337, 340, 347, 351, 357
validation, 217
vapor, 190, 191, 193, 199, 202, 203, 204, 206
variables, 134, 148, 160, 170, 172, 193, 195, 203, 243
variations, 117, 204, 207, 280
vegetables, 347

vegetation, 20, 22, 46, 57, 75, 90, 120, 122, 124, 127, 159, 201, 203, 216, 249, 258, 272, 280, 282, 283, 295, 332
vehicles, 33, 53, 302
ventilation, 145, 153, 175, 183, 184, 357
Vice President, 130
videos, 66, 271
visibility impact, 213, 214
volatile organic compounds, 53
vulnerability, 233

W

war, 108
Washington, 1, 11, 13, 17, 24, 27, 28, 102, 121, 123, 126, 127, 128, 132, 135, 140, 145, 148, 159, 164, 168, 171, 176, 180, 187, 188, 189, 208, 209, 212, 250, 254, 257, 262, 271, 276, 277, 281, 282, 283, 297, 304, 308, 310, 316, 321, 327
waste, 347
water, 37, 52, 61, 77, 81, 83, 88, 102, 191, 208, 210, 211, 221, 255, 256, 257, 258, 259, 279, 283, 286, 297, 315, 324, 332, 347, 351, 358, 362, 364
water quality, 77, 81, 324
water rights, 259
watershed, 67, 174, 294, 301
weakness, 356
White House, 258, 259, 261, 297
wild animals, 279
wilderness, 84, 88, 89, 94, 143, 287, 296
wildfire smoke, vii, viii, 3, 4, 5, 6, 9, 15, 20, 22, 24, 27, 28, 29, 37, 41, 42, 45, 60, 61, 64, 70, 188, 309, 313, 314, 315, 329, 338, 343, 348, 349, 350, 351, 352, 355, 359, 361, 365
wildland, 4, 6, 10, 11, 12, 13, 14, 18, 21, 23, 35, 36, 38, 41, 42, 43, 44, 45, 46, 47, 62, 67, 70, 79, 104, 108, 122, 126, 144, 152, 153, 154, 161, 165, 167, 169, 178, 179,

181, 182, 183, 184, 185, 188, 201, 207, 209, 210, 233, 244, 273, 274, 277, 281, 296, 299, 301, 302, 311, 312
wildland-urban interface, 244, 273, 274, 277, 296, 302
wildlife, 51, 52, 53, 55, 57, 58, 67, 70, 109, 125, 126, 174, 175, 260, 272, 273, 274, 276, 277, 278, 279, 287, 290, 295, 296, 297, 299, 300, 301, 302, 324, 332
wind speed, 119, 203
windows, 46, 119, 133, 134, 147, 152, 161, 162, 163, 167, 169, 172, 180, 183, 184, 185, 186, 187, 315, 336, 337, 338, 339, 340, 341, 350, 351, 352, 357

Wisconsin, 13, 190
witnesses, 3, 6, 7, 8, 9, 19, 20, 21, 23, 71, 74, 98, 102
wood, 30, 55, 81, 83, 112, 150, 185, 315, 332, 337, 340, 351
wood products, 81
woodland, 122, 203
workforce, 162, 256
workload, 31, 174
World Health Organization (WHO), 232
World War I, 108
worry, 51

Related Nova Publications

NATURAL DISASTERS: RESPONSE, RECOVERY AND ASSISTANCE

EDITOR: Sonja Torres

SERIES: Natural Disaster Research, Prediction and Mitigation

BOOK DESCRIPTION: In 2017, four sequential disasters — hurricanes Harvey, Irma, Maria, and the California wildfires — created an unprecedented demand for federal disaster response and recovery resources.

HARDCOVER ISBN: 978-1-53616-407-7
RETAIL PRICE: $230

HURRICANES, WILDFIRES AND FLOODING: DISASTER ASSISTANCE AND CONTRACTING

EDITOR: Lydie Yohan

SERIES: Natural Disaster Research, Prediction and Mitigation

BOOK DESCRIPTION: Chapter 1 provides a short overview of issues Congress may consider in its oversight of the Federal Emergency Management Agency's (FEMA's) federal assistance during the 2017 hurricane season (e.g., Harvey, Irma, and Maria) and other disasters (e.g., fires in California).

HARDCOVER ISBN: 978-1-53616-362-9
RETAIL PRICE: $230

To see a complete list of Nova publications, please visit our website at www.novapublishers.com

Related Nova Publications

EARTHQUAKES AND OTHER EARTH MOVEMENTS

AUTHOR: John Milne

SERIES: Natural Disaster Research, Prediction and Mitigation

BOOK DESCRIPTION: In Earthquakes and Other Earth Movements the author gives a systematic account of various Earth Movements.

HARDCOVER ISBN: 978-1-53616-967-6
RETAIL PRICE: $230

FEMA: EMERGENCY MANAGEMENT, DISASTER CONTRACTING AND GRANTS

EDITOR: Naomi Stanley

SERIES: Natural Disaster Research, Prediction and Mitigation

BOOK DESCRIPTION: Recent hurricanes, wildfires, and flooding have highlighted the challenges the federal government faces in responding effectively to natural disasters. Chapter 1 discusses FEMA's progress and challenges related to disaster resilience, response, recovery, and workforce management.

HARDCOVER ISBN: 978-1-53616-456-5
RETAIL PRICE: $160

To see a complete list of Nova publications, please visit our website at www.novapublishers.com